Word World

Vincent H. Ivory

Janel,
All the best 2 U!!!
Thanks
Vincent H. Ivory

Formatting by Polgarus Studios

Published in the United States by WASP LLC c/o Increase Innertainment LLC, Westerville, Ohio
www.vincentivory.com

ISBN-13: 978-0615822914
ISBN-10: 0615822916

"If quantum mechanics hasn't profoundly shocked you, you haven't understood it yet. Everything we call real is made of things that cannot be regarded as real." – Niels Bohr

Praise
for
Word World

Modern science invents a portal to an Earth-like parallel world ready for colonization—but there's a serious complication: in this new world, utterances become real.

Reading at times like a brainier, Rod Serling–esque version of Stargate's movie/TV franchise, Ivory's first novel.... is a compelling what-if tale that delivers a satisfyingly wise finale.

-Kirkus Reviews

"Sticks and stones may break my bones but words will never hurt me." That's what you think. At first it may seem that Vincent Ivory is making a statement about choosing the things you say carefully. In actuality, Ivory is making a thought provoking allegory of the power of authoritative speech. You will enjoy the sharply developed characters and scenes that weave scientific wonder, political intrigue and greed into this engrossing tale. Parallels to Biblical principles are certain and speaking of parallels, dualities are a common thread that you'll discover as the plot unfolds.

-Darryl Banks, Illustrator & American Comic Artist

Word World *has reawakened my yearnings and spurred my aspirations! After receiving a preview copy of Vincent Ivory's first novel, I was immediately drawn in to its transforming power. Ivory's characters have all the components of human strength, weakness, and frailty. They are flawed, yet redeemable, and he uses them to weave a plot of intrigue, suspense, and ultimately hope. The inhabitants of Dabar reached into my soul, enlightening every remaining dark corner, while exposing every dream, desire, and joy of possibilities available to my unlimited potential. This remarkable story transported me back to the*

Origin and purpose of my spirit, and for that I am grateful. Word World *speaks unspoken promises to unlock the imagination beyond its wildest dreams and as far as you will allow. You will be entertained, challenged, and thrilled by this divinely inspired seed.*

-Marianne Coyne, Author of The Heart of Annie… the Strength of God

Dedication

Big Poppa & My Big Brother –Thee Original Word Masters

Acknowledgements

Special Thanks to:

Le Carmel One
Amy Brown – The Science and Entertainment Exchange
Ryan J. Michney, PhD
Martinelli's
LEGO
Sam
Lavonne Bailey
Sylvain Neuvel
Matt Middleton
Darryl Banks
Ozzie Leon
&
Mr. C

"Twinkle, twinkle little star
Everyone who helped knows who they are!!"

Thanks to all the readers and fans!

P.S. For another great sci-fi read, be sure to check out my buddy Sylvain Neuvel's new book SLEEPING GIANTS! You won't be disappointed.
http://neuvel.net

"The tongue has the power of life and death, and those who love to talk will have to eat their own words."

— *Proverbs 18:21*

Chapter 1

"Finally," a soft voice muttered as the world was quietly ushered into a new era. Unbeknownst to the slumbering masses of Ohio, or the Japanese businessmen finishing up their workday in Tokyo, or the multitude of tourists in Paris who were just now eating breakfast and planning their day of site seeing, life on planet Earth had just undergone a fundamental change in course.

Jerry Jergensen stood over the central control console, the glow of the five separate monitors bathing his face in a pale light. He shook his head as he scanned the data read-out and confirmed what he was seeing. After years of failure he had finally succeeded in creating a device that pulled a parallel world close enough for a stable portal. The portal edges spun and glimmered ever so slightly as it simply stood there like an 8-foot horizontal disc in thin air beckoning to be entered. The two worlds now sitting ever so close. It looked as if you could simply step thru your doorway and then step back in. Everything was running smoothly, just the way it always had in his head. Only this time it was real; this time it worked.

He looked up from the monitors and once more marveled at his work. At the back of the otherwise empty laboratory, thirty feet away from the podium-like control console, Jerry's life's work whirred and spun and bent the known laws of nature. Red laser light bounced and reflected about the lab, a dazzling disco-ball blur of moving lines and dots transitioned from mirror to mirror and this self same light was the only thing holding the portal open. Jerry gulped a few ounces from a bottle of water as he walked closer to the portal. You could see straight thru to the other planet, which looked eerily similar to Earth, its hillside lush and green. Without thinking he threw water from the bottle towards the portal. As some drops fell to the lab floor, others entered the portal and fell into the grass. He paused as if waiting for something unexpected to happen. When nothing did, he quickly donned an environmental suit designed specifically for this day, the day he met with success and then clicked his helmet into place on the suit.

Jerry briefly thought of his wife Candy and his daughter Alexandria as he took a step towards the portal then stepped back.

"Jerry" he said to himself, "We didn't come all this way for nothing! The life insurance is paid up and Candy is still young, she can remarry." And with that selfish rationalization Jerry Jergensen closed his eyes and stepped into the portal. When his foot felt firm ground, he squinted open one eye then the other, like a child expecting to see the boogey man. Instead, he was surrounded by a perfectly blue sky and the green, lush countryside of this new world that looked a lot like his own. He turned around to see the portal at his back and then, coming to his senses he quickly stepped back in to the Earth side. "Whoa!!!" is all that he could say.

Austin Family Practitioners M.D. —Westerville, Ohio, U.S.A.
Earth Date: October 14th, 2014 — 3:39pm EST

The exam room was cold, sterile and had little in the way of decoration with the exception of the anatomy pictures and prescription advertisements that littered the room. The thin exam gown offered no hint of warmth. There weren't even any magazines for Alexandria to flip through while she waited for her doctor. Her bright red painted toenails provided a mix of color against the white floor tiles as she swung her feet back and forth. She had been alone in the room for almost twenty minutes already and the not so soft padding of the exam table wasn't exactly helping with her discomfort. She tried to redistribute her weight, grimacing slightly as she adjusted herself on the exam table.

When she finished fidgeting, she sighed in relief. "That's better."

A few seconds later, a quick knock on the door invaded the silence and boredom of the lonely room as in walked Dr. Kim, Alexandria's physician for more than half her life. "Ms. Jergensen, good to see you again. Didn't I examine you just a few months ago? It's a bit early for another check up."

"I don't need a check up," she said, embarrassed.

Dr. Kim registered the self-conscious tone of her voice and quickly closed the door behind her, ensuring their privacy. She wheeled her laptop workstation over and pulled up her electronic medical chart. "What's going on?" she asked. She sounded concerned, but in a reassuring way. She had always been a great doctor.

"I've been, uh… really uncomfortable lately."

"How so?" she offered with a warm smile. "You're uncomfortable how? You're experiencing pain?"

"Yes."

"Where?"

"In my butt! Well… *not in my butt* but near my butt hole" she shifted in her seat again, wincing.

"Ah," Dr. Kim said. "Lay on your side and let me take a look."

Alexandria tenderly lay down and turned over on the exam table as the doctor put on a blue pair of non-latex gloves. The paper on the exam table crackled as she moved. She tensely stared at the blood pressure machine in front of her as Dr. Kim opened her exam gown then pushed aside her buttocks. In a reflex, Alexandria squeezed. "Relax," the doctor assures her. "Just as I thought," she said after a brief glance. "You can get up and get dressed."

She rolled over and turned back to her. "What is it?"

"Hemorrhoids" the doctor exclaimed.

"Hemorrhoids! You've got to be joking!"

Dr. Kim shakes her head while removing the gloves from her hands.

"But I'm seventeen. Seventeen year-olds don't get hemorrhoids! That's for old people... like you!!!"

Dr. Kim couldn't help but smile. "I could have a picture taken for you if you like. Then you could Snapchat that to all your friends."

"Ugh, this is so not funny."

"I agree. You're right. Thankfully they're very small, although irritating. The good news is that you can get something to treat it over the counter. That way you can use the money you would have used for co-pay to buy a better attitude."

Now it was her turn to smile. Despite her poorly timed joke, Dr. Kim always seemed to have Alexandria's best interests in mind. "Thanks Dr. K. You know just what a girl wants... having to stick her finger up her butt a couple of times a day."

She smiled again as she wrote the name of the medicine she needed on a notepad. "Hey, some people are into that sort of thing."

"Gross," she said not amused at Dr. Kim's humor. "Completely gross. Thanks... for real."

She tore the piece of paper from the pad and handed it to her. "You're welcome," she told her as Alexandria folded it up and stuffed it into her pants pocket lying on the chair. "Call me if you need anything else. And by the way, ... I'm only forty. Forty is the new twenty!"

Prescription Plus Pharmacy — Westerville, Ohio, U.S.A.
Earth Date: October 14th, 2014 — 4:13pm EST

Alexandria's car pulled into the parking lot just a little too quickly, tires squealing as she turned into a spot and slammed on the brakes. The car door opened and the teenage girl pulled herself up out of the unbearably chafing bucket seat. Adjusting herself to as close of her normal stride as she could get, she muttered to herself.

"God, Please…" she said as the store loomed closer and closer. "Let everyone I know or that knows me leave the pharmacy now and don't let anyone come in for three minutes. That's all, just three minutes…"

Eyes darting every which way for any familiar faces, Alexandria went through the automatic sliding doors and headed straight for the aisle that advertised ointments and creams. Scanning the overwhelming inventory of over the counter medication, Alexandria shifted from one foot to the other, trying to find the most comfortable way of standing. Finally, her eyes land on a tube of Preparation H. She grabbed it and whirled around to go pay when all of a sudden—

Ding! The electric bell chimed as the automatic doors slid open once more. Stretching on her tiptoes, Alexandria could feel her eyes bulge in surprise as Peter Troutman walked into the store. Of all the people in the world that needed something from the pharmacy, it had to be Peter. She pivoted around to face a shelf of cheap plastic children's toys, hoping that he wouldn't spot her. But of course, she was too late.

"Hey, Alex!" she heard him call. With her back still turned she shut her eyes and desperately hoped that he would not come over. That he might think it wasn't her after all and move on. "Alex, helloooo…" he said as she heard him shuffling toward her down the aisle. She clutched the tube of hemorrhoid cream behind and shoved her fist into her pocket. *Now watch,* she thought, *a security guard is going to think I'm trying to shoplift and he'll make me empty my pockets in front of everyone.* The way things had been going for her that week; she wouldn't be all that surprised if she were right.

"Hey," she said, forcing herself to turn around and face the boy who had teased her since elementary school. They had never gotten along and yet somehow they always ended up stuck in the same class together.

"What are you doing here?"

"I, uh, I'm just—"

"No wait! Don't tell me! Let me guess! STD? Again?"

"No, *P.I.T.A.*, unlike your sister I know how to keep my legs closed."

He sneered at her, not because of the jab at his sister, but because of the way she said his name. She had been taunting him with it ever since they were in fifth grade, and it was still like grating on his ears. Whenever he pissed her off she called him *P.I.T.A* instead of Peter.

Of course Peter had made fun of her in their early school years as most young boys do. The years Alexandria called '*her not so cute days.*' And so, in fifth grade, she came up with an acronym for him and decided it was time for a little payback. She changed his name to *P.I.T.A*, which meant only one thing: *Pain In The Ass.*

"That's not my name," he told her, his jaw clenched. "It's Peter. You know that."

"Whatever, *P.I.T.A.*" she grinned and walked past him, praying that he wouldn't follow her to the checkout counter. Peter rolled his eyes and continued down the aisle to the beverage refrigerator. She rushed quickly to the register. She paid the clerk and didn't bother to wait for the old woman to make change as she bagged the Preparation H and handed it over to her. Alexandria grabbed it and all but ran out the door.

"The old woman hollered, "Hey you forgot your change!" "Put it in the little tray" she told her, yelling back.

She walked back to her car, wincing as the hemorrhoids whined and bristled at the movement, completely oblivious to the irony of her ailment and her running into Peter. She had been making fun of Peter, calling him a name that meant *Pain In The Ass*, for years. Now, at the tender young age of seventeen, she herself had developed her very own case of *P.I.T.A.*

Coincidence? Perhaps, or was it really?

What Alexandria and the rest of the world would soon discover was that there are connections in the universe. Connections that mean much more than people might think. Strings of cause and effect made from strands too small for most to see. Like most human beings, Alexandria was still completely unaware that her words were much more than simple verbal tools of communication. But, not for long.

Troyconics Astrophysics — Columbus, Ohio, U.S.A.
Earth Date: October 14th, 2014 — 5:53pm EST

Ben Levin sat in the Troyconics break room, a canned energy drink in one hand and a leg propped up on another chair. He sipped his drink while he watched the old action movie that played on the employee television. The sound was muted, but he stared at the quick-cut fight scene all the same.

There were a few other employees on break, but they were either working through their meals or had their eyes glued to screens and they're fingers dancing across keyboards. The handful of work-obsessed physicists and engineers made the room feel more like a library than a place to take a half hour break. Not for Ben though. For him, taking a break meant taking a break. And if that meant staring at a silent television, then that's what he was going to do.

It wasn't until Jerry Jergensen shuffled into the break room that Ben looked away from the TV. He had worked with Jerry at Troyconics for years, and considered him a friend, but he rarely saw him in person. For the most part, Jerry was sequestered in his lab where he continued work on some mystery project that everyone else in the R&D department had no knowledge of let alone security clearance. This was the first time Ben had spotted the man in months.

Athletic, tall and in an extremely good shape for a man in his forties, Jerry trudged across the linoleum floor with a prideful, almost selfish confidence. A few more grey hairs had sprouted from his locks but what had not changed was Jerry's signature outfit. An expensive black mock neck with grey dress slacks covering his tall frame and black Kenneth Cole air sole shoes to complete his ensemble. "He's alive, he's alive!" Ben exclaimed, ever the jokester. "Uhhh... didn't you have on that same outfit the last time I saw you, like three months ago?" Ben continues.

Jerry looks around annoyed by the worn out office joke about whether he has twenty of the same outfits or just one. He grabs Ben in a headlock, ensuring that his armpits are near Ben's nose, "Yes it sure is, get a whiff of my ninety day funk!"

"*Shhhhh!*" one of the other employees said over the screen of his laptop before taking another bite of his sandwich. Ben gave the engineer a dirty look and got up, following Jerry over to the coffee machine. "Shhhesh yourself. Gimme a break! This is the break room you know?" "The rest rooms are down the hall for constipated people like yourself!" They both grab their drinks and sit at a table by themselves.

"What's going on, Man?" he said, remembering to keep quiet. "Long time no see."

Jerry nods as a takes a sip of his hot chocolate. "Please tell me you are not going to drink coffee and an energy drink at the same time?" Ben takes a sip of both beverages. "Oookay, that answers that." he said, as he returned to Ben's question. "I've been, uh… working."

"Right, on your mystery project. You know, if you're ever willing to talk about what you're working on, I'd love to hear more about it."

A smile spread across Jerry's face, a bright slash of excitement and satisfaction. "You know what, Ben? I think I would like to talk to you about it." "But first I'd like to talk to you about something else" he leaned in, whispering to Ben. Whether he whispered so he wouldn't disturb the others or so no one could eavesdrop was unclear.

"So," Jerry started, "there's this guy—a scientist in Japan—who's been studying the effects of words, thoughts, and music on water."

"You mean the crystal guy? The snow flake dude?"

"If you want to call him a dude, sure. You've heard of him?"

"I've heard of him. I don't know if I believe it, but I've heard of him."

"Alright, well let's say you *do* believe him—"

"Oh boy, so we're really going there huh? To Jerry's world? Word world?" Ben grinned playfully. Teasing Jerry for his obsession with linguistics.

"Don't be an S.A." Jerry smirked.

"You got it, Esé," he shot back in a faux Spanish accent. "What is it with you, Jerry? You won't cuss, but you'll say the initials? I know S.A. means Smart Ass when you say it."

Jerry leaned back in his chair and shrugs. "Well, technically"ass" isn't really cussing. Ass is in The Bible."

"Are you serious?" Ben asks in unbelief.

"Sure. Ass, asses… is in The Bible like a hundred times."

As Jerry continues with his *assplanation*, a woman—a technician from the applied physics department—walks into the break room and heads for the fridge. Ben, who was a little more than half Jerry's age, couldn't help following her appealing figure with his eyes. "Somehow, I don't think The Bible is talking about the same kind of asses I'm talking about."

"Can you stop?" Jerry asked, snapping Ben's attention back to him. "Use your other head for just a few minutes."

"Right. Sorry. I forgot. Leave ass world go back to Jerry's world right now."

Jerry ignored the teasing and tried to salvage the conversation, picking back up where they had been sidetracked. "Anyway. So this doctor says that *words* affected the water and the crystal shapes. Different words did different things to them."

"And?"

"And, the human body is made up of like sixty percent water. Doesn't that make you wonder?"

"Wonder what?"

"Wonder what might happen if people constantly just spoke positive words over their body? How would that affect the water molecules in their body? How would it affect them? Their health? Maybe it would have a similar—"

"Ooooookay… Since you already brought up The Bible, let me ask three questions. Are you about to go Holy Roller on me?"

"No."

"Name it and claim it?"

"Nope.

"Believe it and receive it?"

"Not a chance."

Dinner in hand, the attractive physics tech walks past their table again. Ben's eyes were drawn to her once more as if through magnetic attraction and he shook his head with appreciation. "Reach out and take it…"

"That's four questions."

"Oh, that wasn't a question. I want her."

Jerry turned around and looked at the woman for the first time. Then he snapped his head back to Ben, who was still staring. "You need help. Come on, focus."

"Maybe she's weak-minded," Ben trailed off, still watching the woman as she sat at the table. "Perhaps I could use The Force on her?" He lifted a hand—the one not clutching the energy drink—and contorted it into some sort of psychic claw, staring the woman down and trying to will her back to him with nothing but his mind.

Rolling his eyes, Jerry got up and grabbed Ben by the arm. "Come on. Let's just go to my lab.. adolescent!"

He led the younger physicist out of the break room and took him through the labyrinth-like corridors of Troyconics Astrophysics. The hallways were white and sterile, lit with harsh fluorescent bulbs that simulated constant daylight. They made their way past various windows in which men and women in lab coats could be seen operating machinery and recording data.

Finally, the walls become simple grey concrete as they came to a service elevator and Jerry swiped his I.D. tag over the sensor before pressing the "UP" button. As they waited for the elevator car to arrive, Ben continued to ask Jerry about his theory. "So if words can affect things the way you say, how come I can't just talk up a million dollars?"

"It doesn't work that way. It's not like a parrot, where you just way whatever. It's like gravity, there are laws that govern this stuff."

"I thought you said you weren't going to pull the Holy Roller act on me. This sounds like faith stuff to me." My mom's Evangelical. Speaks in tongues, falls under"The Power," the whole bit.

"Well it is in a way," Jerry admitted as the elevator *dinged* loudly and the doors slid open. The two of them stepped into the large box of the elevator car and Jerry swiped his I.D. tag again before entering a security pass code and the elevator begins to *descend*. "Uhh, I didn't know this elevator went down? Should I be concerned?" Ben asks slightly worried. Jerry smiles, ignores his inquiry and continues his conversation as they began a long descent. "It's what you believe and some other pieces along with it. Your words... they're like a creative force. The words you believe in. They actually do something. It's like a world on top of another world but the other world's slightly different. Different laws, different principles... I don't know everything. I haven't figured it all out yet. I only know a little bit."

Ben stared at Jerry with musings of confusion. "Jerry, just what exactly are we talking about here?"

"I'm not exactly sure yet," Jerry told him, beaming.

The elevator finally settled and the doors slid open. Ben didn't know how deep underground they were, but the air was cold and dank. Stepping out of the elevator, Jerry led him in the only direction they could go: forward down the concrete tube that stretched away from the elevator and ended at a large blast door. The thick steel barrier was at least thirty feet wide and twenty feet tall, it's seam zippered with the black and yellow paint of warning stripes.

"This is where you work all day?" Ben asked a bit overwhelmed.

Jerry ignored his wonder. "Ben, everything I'm going to tell you now is four levels above your clearance level. I've already put in a request to have your clearance level upgraded. I'm going to want you on the team."

"What team?" Ben asked as they approached the humongous door.

Tapping an eight-digit access code into the control panel on the tunnel wall, Jerry swiped his I.D. badge one last time then spoke his name. "Jerry Jergensen."

There was a deep rumble as motors worked to split the door down the middle, each half sliding into the curved concrete walls. Jerry turned back to Ben as he stepped backward into his subterranean laboratory. He smiled and waved for Ben to follow him.

Ben took a step over the grooved threshold and entered the huge open space that was Jerry's lab. At least a hundred feet high and several hundred feet across, their footsteps echoed as they walked across the vast floor. Large computer servers and mainframes were stacked in clusters, with anaconda-sized cables connecting them and twisting along the floor. There were several different stations with computer consoles built into them; everything had clearly been custom built.

In the center of the giant room, was what Ben could only assume was Jerry's mystery project. At first glance, it looked like some sort of carnival ride. Two metal arms branched out from three central motors built into the ceiling. Each arm had yet another motor at the end, with four smaller arms branching out from those. The end result was something that resembled a giant mobile, and from each of the smaller arms, hung two dozen of the largest mirrors Ben had ever seen—at least twenty-five feet tall.

"Whoa…" he muttered as he looked at the fractured reflections of twenty-four differently angled mirrors.

It was like looking through a mirage. In some he could see himself. In others he could see Jerry. In others there was just empty space. "What is it?" he asked.

"I call it a Dimension Mirror. I was thinking if you could figure out a way to bend laser light so that it's constantly—" he stopped, getting ahead of himself. "You know what? Why don't I just show you what it is?"

He walked Ben over to a podium of computer consoles that faced the mirror mobile dead center. Then, as casually as one might pop quarters into a pinball machine, Jerry flipped a switch and the five monitors in front of them lit up. He began tapping at the touch-screens and inputting commands.

There were several loud, mechanical *clanks* and *whirs* as the three central motors, which were positioned on the ceiling as the points of an equilateral triangle began to spin. The arms twisted around with them, and the smaller ones that split off from those soon joined in the dance. The mirrors spun in conflicting orbits, getting faster and faster and yet somehow they never smashed into one another.

Ben stared in amazement as the giant reflective panels reached their optimal speed—Ben guessed they had to have been spinning at least 250rpm—which was *fast* when you were working on a scale this large. The

15

reflections became a blur.

"Oh, I almost forgot!" Jerry yelled over the whirring grind of the machinery. "You're going to want to put these on!" He raised his hand and Ben saw a pair of dark-lensed safety goggles dangling from his finger. He asked no questions, he just grabbed the goggles and slid them over his eyes. Jerry put on his own pair and then went back to tapping on the monitors.

Looking away from the spinning blur in front of them, Ben looked over his colleague's shoulder. From what he could make out from the highly intricate data read out on the screens, the mirrors were rotating according to an algorithm Jerry had programmed into the computer. He couldn't be sure without going through the coding, but Ben was willing to bet that the mirror mobile—the "Dimensional Mirror", as Jerry had called it—could spin forever and never repeat the same exact positions twice.

"Activating laser light now," Jerry said.

"What?" Ben asked. He could barely hear anything over the noise.

Jerry either didn't hear him, or he simply ignored the question, but before Ben knew what was happening, Jerry reached out and hit one last button and the giant cave of a lab exploded with red laser light, shooting out in all directions. Squinting through the dark lenses of the goggles, Ben looked into the spinning mirrors.

The laser had originated from one spot in the center of the triangle made by the central motors. It shone straight down through the middle of the frenzied mobile where it struck the top of a mirrored sphere and reflected out in all directions. Those fractured rays of laser then bounced off the constantly moving mirrors, changing course every fraction of a second as the mirrors shifted and spun around it.

The effect was a dazzling amorphous blur of twinkling red sparks that somehow seemed to boil and bubble into a sphere of radiance. It grew in size, expanding outward like a simulated scaled-down model of the universe. When it finally reached the edges of the mirrors it flattened on all sides, crackling with energy and light. Pulses of fragmented laser particles stretched like webs across the face of the Dimensional Mirror, then gaped open with a clarity that made Ben's jaw drop.

Beyond the energy field, Ben could see a rolling valley, lush with green vegetation. A vibrant blue sky with no clouds in sight. He could feel the energy of the entire contraption against his skin as he looked into the open portal. He put an arm on Jerry, steadying himself. "Don't tell me you built your own Stargate!"

"Kinda!" Jerry cried, a humongous grin plastered on his face as he removed his goggles. "You're the first person I've shown it to!"

"How long have you been hiding this?"

"About six months!"

"Six months! And you didn't tell me? You didn't tell anybody?"

"I couldn't share it! Not until now. But I always planned to have you as part of the team. And when I say you're the first person I've told I mean it. You're the first person outside of the construction team that has any idea what I'm doing here, and even they only have educated guesses. I haven't even told Calvin about this."

"Uh, Calvin is like… the boss?"

"I know."

"And do you really think it's a good idea to not to tell your boss about *this*?"

"I will tell him. Soon. I just wanted to be careful. I don't want bureaucratic policies and red tape to get in the way of the progress I've made here. Anyway, I'm telling you first. Please let that count for something and help me keep it a secret a little while longer.

Ben stared at Jerry, then turned back to the swirling but stable portal. "Okay. I'll keep your secret… for now. But only because I'm more interested in finding out more about this thing than I am in kicking your double crossing butt right now."

The Sanctuary of Us — The Planet Dabar
Date & Time Unknown

As Jerry Jergensen and Ben Levin stared into the portal at the new world, the new world seemed to stare back at them. The portal had opened onto the grassy slope of a glacier-cut valley. It was bursting with flora and fauna of bright colors that seemed even more vibrant thanks to the crispness of the atmosphere.

A few miles away, almost completely hidden from view behind a copse of trees, stood a structure that a human might confuse for a medieval castle. Large and palatial, it's stone walls were both sturdy and beautiful, carved down to the smallest detail. Smoothed blocks of dark grey rocks, assembled together in a stunning display of engineering and creativity.

Atop the highest of the many open terraces and balconies, a figure stood at an intricately designed railing, staring out over the valley floor. Though a human eye would not have been able to spot it at such a distance, the humanoid creature at the edge of the balcony watched as the portal shimmered and finally closed without a trace.

The ThreeofUs stared out at the spot where the door between worlds had appeared, his gaze passively pensive. If he were simply human he would pass for seventeen, however he... they had become so much more than simple age could define. He was something else and knew it. He sported white hair and his piercing blue eyes conveyed wisdom far beyond seventeen or even forty years of age. Though clothed apparently in humanity it was only an outfit, a window dressing of their true being. He like the others with them had been set free from and refused to acknowledge the weakness of individuality and now stood in the immense strength, knowledge and Oneness of *The Us*.

Joined now by the OneofUs, they both stood and stared from the balcony. "How many have joined us?" he asked of OneofUs, his voice ethereal, somehow close and distant at the same time.

"Too many," OneofUs replied. "...And then not enough."

"Is that to say we have quantity and not quality?"

"Precisely. Some have come to learn. Most have come for selfish reasons... Although they have not realized what they will become yet. Selfish..."

The ThreeofUs turned back to the railing and looked out at where the portal had opened. "Such is the way of all who begin for intrigue and not purpose."

"Perhaps they will change," OneofUs offered. "As we did. By receiving that which has made us different?"

"We grew better. Not that we knew better. The *death* of ones close to us... is a fast teacher..." The OneofUs goes back inside, leaving the ThreeofUs to contemplate the new visitors. His thoughts rolled back to his first visit to Dabar. Things were different then. He was different then. "Yes..." he said aloud. "A fast teacher indeed..."

Planet Dabar
Date & Time Unknown

There had been twelve of them that day. The ThreeofUs remembered it as if it had been yesterday. The day when he and his team discovered Dabar. The memory was clear and crisp. He could even still remember his Earth name: Darren Abercrombie.

His team and colleagues, some physicists, some scientists, some ex-military but all specialists in their field had made their first contact on Dabar. By virtue of happenstance, they had some how discovered a way to bridge their two planets without knowing the other existed. Yet being scientist at heart, they stepped forward through the portal created accidentally during their tachyon field test. After first sending in a probe, Darren and his team walked out into the open air of Dabar, each of them protected in an airtight, oxygen filled Environ Suit.

"Wow!" he had cried as he looked around at the lush landscape. "What is this place? It looks so much like Earth. Perhaps the air is even breathable."

His colleague—Milisa Tanwa, —checked an atmospheric scanning device that she held in one hand. "It doesn't appear… to be…"

"What?" Darren urged her. "What is it?"

"It's weird. The air was registering as having zero oxygen… But when you said,"hopefully the air is breathable…"

"What?"

"It's almost as if it changed when you said it!"

"Check your scanner, Tanwa!" yelled Chet, the teams other environmental engineer. "Air compositions don't just change like that. Either your scanner's bad or you need to check the air mixture in your—"

"I know what I saw," she cuts him off.

"Okay everyone, let's calm down," Darren quelled the tension. "This place doesn't appear to be too different than our world. The vegetation and plant life seem almost identical. If there's not oxygen on this planet, I'll be totally shocked."

"Are you sure this is even a planet?" Ferguson asked.

"A parallel planet perhaps. World. Dimension. Whatever it is, it's connected to Earth or we've connected it to Earth, …somehow. It looks like Earth. Who knows, either way we've ripped something open in time or space? Milisa, is the atmosphere still registering as breathable?"

"Yep! Sure is, boss," she reported.

"Alright then," Darren said, as he unlocked the clamps on his helmet.

"I wouldn't do that if I were you…" Brett exclaims with caution

"No, boss!" Chet cried, as he reached for Darren's helmet to put it back on. But their fear was unfounded. Darren took a deep breath and let it out, smiling.

"Wow," he said. "That's fresh. It's okay. You can take off your helmets."

Ferguson stood in the unnaturally green grass, shaking his head, the face behind the shield of his helmet scared and panicked. "Hell!" he cried. "I'm not! I could die!"

Then, suddenly, there was a monotone *beep* from off to the left and all twelve heads swiveled to look for the source of the sound. There, attached to the top of a four-foot pole, was a small metal box. There was a small black screen centered on each face of the space gray cube, giving it the look of some kind of futuristic parking meter. In each of the screens, a red LED display lit up the number *6*.

Then the screen blinked and the number changed: *5*.

Then: *4*.

"What the hell is that thing?" Ferguson shouted, the sound both muffled by their helmets and amplified by their comm links. He pulled the precautionary sidearm that was strapped to his thigh out of its holster and looked around, paranoid. He aimed his the pistol at the box as it continued its countdown. "Is that a bomb?"

2.

"Don't shoot it," Darren urged him. "Just stay calm. We don't know what it is."

1.

"It's a bomb! We're all going to die!"

0.

Darren heard a booming surge of static in his ear and watched as Ferguson dropped to his knees, then face first to the ground. Darren ran over to him, screaming. "Ferguson! Ferguson! Do you copy" He ran over, hoping to find a pulse, but he could tell before he even reached the body: Ferguson was dead.

"Darren, what's going on?" Milisa yelled. She ran over and knelt beside Darren. Ferguson laid face down before them, steam rising off his body. "Is he dead? Let's turn him over—" she reached out to help flip him onto his back, but as soon as she touched him she yanked her arm back, yelping. "Ow! His suit is hotter than hell!"

Beep! The screens on the metal box reset once more, its red LED's making the shape of a 6. Again, it began to count down.

Making eye contact with her through her helmet visor, Darren could see that she was sweating profusely. Her hair was pasted to her forehead in wet clumps. "Is anyone else burning up? I feel like my suit is on fire—"

Milisa looked from the device back to Darren, her eyes wide with fear. "Was that me? Did I do that? Oh God, am I going to—"

"Stop it!" Darren yelled as the seconds ticked by. "You're going to be okay!"

"Oh God, Darren it's so hot! I can't take it! I feel like I'm going to burst into flames!"

The box hit zero again and suddenly Milisa was consumed with sudden flames. Darren stumbled back, feeling the heat through the thickness of his suit. She was screaming and flailing, as she burned alive.

"Help her!" someone shouted. "What the hell is this place?" cried another voice. *Beep!* Darren heard and saw the box reset its countdown.

"What the hell is this, Darren?" another team member yelled—either Morgan or Brandy, Darren could no longer be sure which anymore— you're going to get us killed!"

"Quiet!" Darren tried to tell them. "Everybody, be quiet!"

"I didn't sign up to die," Morgan told him.

"Listen to me, Darren tried desperately. Nobody else say anything else

22

about dying. Just stop talking—"

"No, Morgan is right!" the team physician Brandy shouted over him. "We have to get out of here!" Brett then panics and shouts, "We're *all going to die!*"

"No—" Darren tried, but it was too late. The remaining team members were all yelling over him now. Yelling over one another.

"Shut up, Brett!" Tanwa shouted. "Alright I take it back!"

"What do we do?"

"My suit feels tight… I can't breath!"

"Shut up!"

"It's the suits!"

Morgan grabs Darren by the collar of his suit and screams, "DO SOMETHING!!"

Darren tried once more to get through to them, but to no avail. "Everyone please be quiet! It's your words! Don't you understand it's your words—"

The screens on the box hit *0*, and suddenly there was nothing but silence. Darren looked around, still breathing the fresh oxygen rich air. All but four of his colleagues lay about him, all of them dead.

Jergensen Residence — Westerville, Ohio, U.S.A.
Earth Date: October 15th, 2014 — 4:28pm EST

Alexandria's bedroom floor was a sea of cast aside clothes. Her bed floated amidst the waves of cotton and silk and she lay across it like a shipwrecked castaway, deserted on an island. Turning over on her back, she rested her feet and legs against the wall, her cell phone glued to her ear as she stared up at the posters covering her ceiling.

"You can't tell anyone, I'm serious!" Alexandria said sternly into the receiver. "If you do we are done for life."

"*Girl, you know I wouldn't do that,*" said the voice of her friend Kendra on the other line. Best friends since grade school, Alexandria told Kendra everything. Even if it meant dealing with Kendra's boisterous attitude when all she wanted to do was get some sympathy. "*But seriously. Hemorrhoids?*"

"They should be gone in a couple of weeks."

"*So what's it like?*" Kendra asked.

"You don't want to know."

"*It hurts?*"

"Like I said: you don't want to know. I had to go to the pharmacy and get Preparation H like some kind of geriatric. Oh, and get this: I'm trying to get in and out of the store without anyone seeing me and guess who walks in while I'm on my way to the register with this tube of hemorrhoid cream?"

"*No! Not Peter!*"

"Yes! I was so embarrassed, I tried to hide, but he walked straight up to me and started giving me crap. He was all like: 'what are you in here for an STD prescription or something?'"

"*No he didn't.*"

"Yes he did! I told him he must have been confusing me with his sister."

"*Or his mom. Girl she is kinda* out there, *know what I'm saying? I saw her at the club, can you imagine? Fifty-two acting like she twenty-two. Really? She need to go home and watch Desperate Housewives or something.*"

Alexandria burst out laughing, ignoring the discomfort it caused as she bounced on the mattress. "Girl, you are crazy! That show isn't even on anymore."

"*That's what I'm saying! She need to not be at the club no more!!! Thinking she a bobcat or something.*"

"Bobcat?" Alexandria cackled. "You mean cougar?"

"*Girl, cougar, cheetah. Snow Leopard. Lion. Mountain lion, whatever. She need to quit, that's all.*"

There was a *beep!* in Alexandria's ear and she took the phone away from her ear to look at the screen. There was a call waiting. *Dad,* the screen read. "Hey, hold on, it's my dad," Alexandria told Kendra, before putting her on hold and taking her father's call. "Hi Daddy!"

"*Hi Baby,*" Jerry Jergensen's voice spoke in her ear. "*What are you up to?*"

"Just on the phone with Kendra," she told him, twirling her hair unconsciously.

"*Okay Sweetheart. Just tell Mom I'm on my way home. I tried to call but I only got her voicemail, so—*"

"So you hung up. Really Dad? Leaving a voice mail is not that serious."

"*I know, you're right. I love you, Lexi. I'll see you guys soon.*"

She smiled at the pet name. "Love you too, Daddy. Bye."

Pausing only to switch back to Kendra's conversation, Alexandria picked up her story where she had left off. "Kendra? Hey, yeah, that was my dad. So anyway, I'm trying to pay for this ointment and get outta there, and who walks in? *P.I.T.A.!*"

"*Girl, how many times I gotta tell you to stop calling that boy that. Don't stoop to his level. You put bad stuff like that into the universe and it gets bounced right back at you. Who knows? That might be how you got those 'roids.*"

"Kendra!" Alexandria laughed at the idea. "Whose side are you on?"

The Jergensen home was a nice split-level ranch in a small suburb outside of Columbus, Ohio nestled at the end of a residential cul-de-sac. Jerry pulled into the driveway, got out of his car and walked up the back porch steps, whistling as he climbed them. He entered the back door and walked into the kitchen with a handful of files and other papers.

A pungent aroma filled his nostrils as he closed the door behind. The room was filled with flowers. In fact, almost every open surface was covered with some kind of vase or pot. The colors were brilliant and clashing. Jerry breathed the scent in as he smiled at his wife, Candy, who was washing broccoli over the kitchen sink.

"What's for dinner?" he asked, trying to be cute. "Flowers?"

"Ahhh," she said, not bothering to look up from her task. "I suppose a bad joke is better than no 'hello' at all."

"Hello," he offered, too little too late.

"Hi Jerry. Why are you home so early? What's the matter?"

"Nothing is the matter," he told her, walking around the island counter and placing a gentle kiss on the smooth skin of the side of her face. She looked inquisitive still, but her lips couldn't help curving into a slight smile. "Someone's being attentive all of a sudden… What's going on?"

Jerry gaped at her, a little offended. "Candy, that's cold. You know I've kissed you at least a twice this month."

"That maybe, but you tongue kiss your work all day, everyday. It's a good thing I'm not the jealous type." She moved the broccoli over to a cutting board and picked up a vegetable knife. Chopping through the green trunks of the broccoli, she winked at him to let him know she was teasing. "So what happened? Your girlfriend at work stop putting out?"

Jerry smiled at her and headed for the hallway, smacking her on the butt as he walked past. She jumped at the sudden squeeze he gave her left buttock and giggled. "Oh! Frisky too. Next thing you know you'll be asking for sex!"

The thought lingered in Jerry's head as he made his way up the stairs toward Alexandria's room. Her door was ajar, light spilling out into the dimly lit corridor. He walked up to it slowly, pausing to eavesdrop on his daughter for a moment.

He wasn't proud of it, but he couldn't help it. These days it was the only way to know what was going on in her life.

Though he couldn't hear her this far from the phone, he assumed that Alexandria was listening to Kendra. After all, the two of them had been practically inseparable for years. Even when they weren't in the same room, they were talking on their smart phones or texting each other.

And Jerry was right. At that very moment Alexandria was listening to Kendra's voice on the other end of the phone. *"The truth is, you* like *Peter. That's why he bothers you so much, Girl!"*

"That is so not true."

"I mean you gotta admit it, Peter is cute. You know he is…"

"He may be, but he still acts the same way that he did when we were fifteen."

"He's a boy! He can't help it. But a cute boy…!"

Jerry finally gave up on his attempt to gather any meaningful intel and knocked on the door gently. "Hi Baby, I'm home."

"Hey Dad," she called to him from behind the closed door. Kendra let me call you back."

"Girl bye!"

She hung up the phone as he entered her messy, teenager's bedroom. "Hey," he smiled at her as she rolled over onto her stomach and looked up at him, her feet bouncing behind her.

"Why are you home?" she asked as she dropped her phone onto the bed.

"Because I live here. Can't a man come home to see his family without being interrogated?"

"Not without ulterior motives, Dad."

"Et tu Brute?"

"What?"

Jerry's smile flipped downward in a concerned frown. "They don't teach you'll Julius Caesar in school anymore?"

Alexandria rolled her eyes in the back of her head. "Is that a Cro-Magnon man from the stone ages or something? You know, back when you were young?"

"If I weren't young would I do this!" he cried, doing his best to laugh off her hurtful comment. Taking a few running steps, he launched himself into the air and twisted sideways. He landed on the mattress with a springy creak and bounced next to his daughter, wrapping her in a loving hug. "You and your mom are so cold today."

"No, we're just smart enough to know that when you leave work early like this it means you want something. So what it is?"

"I just want to spend time with you two. Besides, there's something I'd like to talk to you both about."

"I knew it," Candy interjected, appearing in the doorway and leaning one shoulder against the frame.

"Me too," Alexandria chimed in, shrugging Jerry's arm off of her shoulder.

"Okay, what is it?" Candy asked. "Is the electric bill over three hundred dollars again and you want us to turn out the lights? No, it must be the water bill this time from Lexi's long showers."

"Stop," Jerry told her. "It has something to do with my work."

"Dad! Did you get fired?"

"What's going on, Jerry?"

"No, no, no," he explained, waving their concerns away with both hands as he sat up on Alexandria's bed. "I'll tell you guys at dinner. That way we can really sit down and *talk* about it. After all, you're the ones who always say I never talk to you about my work," he said to Candy. "I just need your guys' help with something."

They watched him from both sides as he leaned over and kissed Alexandria on the crown of the head. Then he left the room, whistling again. On his way out, he gave Candy's butt another smack and whispered in Candy's ear. "I'll take you up on that sex now."

Candy's eyes went wide with surprise. At the same time, Alexandria covered her ears with her hands. "I heard that! Ugh, I'm going to the mall. I don't want to be around when you guys come out of the room with that look on your faces as if you've done nothing."

"Just be back for dinner…" Candy trailed off, smiling as she followed Jerry down the hall and into their bedroom.

5:32pm EST

Later, Jerry and his wife found themselves lying in their king-size bed, their bare limbs intertwined comfortably and familiarly. Their bodies had always fit together in the strangest ways, like some sort of contortionist duo. Now they laid together, basking in their post-coital bliss.

Candy groaned with content as she ran her fingers across Jerry's chest. "So, did you come home for sex and the talk was an excuse? Or did you come home to talk and the sex is an excuse?"

"Hey, can you cut me some slack? Can't I do anything without being suspected of ulterior motives?"

"Sure! Either way, that was one of the best 'excuses' I've heard in a while," she murmured in his ear before nibbling it and getting up to use the bathroom. They both laughed as she disappeared into the other room and shut the door.

There was a moment of quiet as Jerry closed his eyes. At least until he heard the low bass-rumble of a fart from the other side of the door. "Sorry," Candy called out from the bathroom. "It slipped."

He couldn't help but be taken out of the mood. "Well, I guess that takes care of round two…" he said, rolling out of bed.

6:54pm EST

"You haven't lost your touch in the bedroom or the kitchen," Jerry announced with delight as he finished up his dinner.

"I'm right here! Please cease with the gross parent sex talk." Lexi declared as if repulsed by the thought of it.

"Lexi, please!" Candy smirked while grabbing another roll. "You're seventeen now." Jerry interjected, "Yes and hopefully, you are STILL not having sex." Lexi's eyes roll again in the back of her head as she swallows her food, "I'm not Dad, don't worry."

They had just sat down to dinner—Alexandria in the new shirt she had picked up at the mall – Candy then decided to change the subject.

"So, are you going to tell us? What's the news? Or did you just come home for sex after all?"

"Mom!" Alexandria said in disgust, dropping her fork to her plate. "Please, I'm trying to eat!"

Jerry ignored the teenager's over-exaggerated repulsion and looked across the table at his wife. "Well. I, uh, I just wanted to let you know that I'm going to be taking a little trip pretty soon."

"A business trip?" Alexandria asked.

"And just where are they sending you?"

"Well. I'm not exactly sure, but let's just say I'm going to be going into space."

"Space?" Candy asked, her voice filled to the brim with skepticism.

"That's why I wanted the sex," he grinned playfully. "You know it takes a long time to go up to space and come back. I figured I needed some now since I'll be gone for a bit."

"Dad!"

"Sorry, Lexi," he laughed. "Okay, okay, seriously now. I wanted to tell you both that I've made a… a pretty big discovery at work."

"Really?" Candy asked, genuinely interested. "How big are we talking?"

"Big," Jerry confirmed with a solemn bob of his head. "Life changing! And I mean that in every sense of those words. This discovery could change life on this planet."

Alexandria and Candy looked at each other from across the table; both of them wondering what Jerry could be talking about.

"Anyway," Jerry continued. "I wanted to let you know that and I still need to lock down some of the major details before I announce what I've discovered. Until then I may be spending more time at the lab then I usually—"

"There it is," Candy said, disappointed.

"What?" Jerry asked.

"That's why you came home, Dad. You don't have to pretend, it's so obvious."

"No, I wanted to tell you about what I found—"

"Except you can't tell us what you found," Candy told him. "All you told us is that you're going to be spending more time with your other girlfriend - the laboratory."

"I know. I'm sorry. But it would only be for a couple of months or so. Please guys, work with me."

There were several prolonged seconds of silence as Jerry's wife and daughter thought this over, looking at each other. Finally Alexandria smiled. "I'm cool! Just give me money. You know, pay me off for all of the psychological damage you're doing by not being here. After all, shopping heals all wounds."

The table erupted as all three family members burst out laughing. All of their differences aside, they loved each other dearly. They were a family—a team—and they could still laugh together as one. "I want to go to Europe," Candy said candidly.

"Europe?" Jerry asked, caught off guard. "Why Europe?"

"Jerry! You're going to outer space! I want to go to Europe. I think that's a fair deal." And with that, she stood up from her chair, gathering up the now-empty plates and walked back into the kitchen with them.

"See Dad?" Alexandria pointed out, excusing herself from the table and walking over to Jerry. "You got off easy with me. Just a *few hundred dollars*. Mom is going to cost you about ten grand when your little trip to the moon is all said and done."

She bent down and kissed him on the cheek. He couldn't help but smile. "Why do I feel like I've already been taken on a trip?"

"That's because you have, Daddy. Didn't they tell you? Outer space is expensive."

Then she walked out of the room, leaving Jerry sitting alone at the head of the table, wondering how he had let that conversation get away from him so quickly. He chuckled to himself, shaking his head. "Europe..." he muttered, baffled.

Chapter 2

The rolling crests and ridges of the Palermo province of Sicily were a strange hybrid of arid desert and fertile vale. Scrubby bushes and olive trees shared the soil of the Italian landscape as the small village of Corleone loomed above, positioned at the top of a gradual hill.

Removed from the village, Bernard Maychoff's palatial estate, sits on the North-west side of the hill, and on especially clear days, when one stood on the villa's highest terrace, they could just make out the misty, placid waters of the Mediterranean Sea. But presently, the lights in the windows of this astute mansion are the only things that permeate the evening darkness on the hillside.

Inside, Mr. Maychoff sat in a rustic leather chair in his home office, surrounded by highly polished cherry wood and other elegant décor. A painting of a nude woman with a slight pout, stares back at him as he converses on the phone with the director of one of the many astrophysics companies he owns. Calvin Madison, the director of Troyconics Astrophysics and Jerry's 'boss,' is on the other line attempting to answer Mr. Maychoff's many inquiries.

"So? How are things progressing in the research and development department?" he asked into the custom made ivory telephone he held to his ear.

"*Very well,*" the voice on the other end informed him. "*Jerry seems to be progressing very nicely,*" Calvin told him.

"Seems?" Maychoff said unconvinced. "So once again you have not actually *seen* his most recent work?"

"*Mr. Maychoff, as I informed you before, Jerry is very independent and we agreed to give him a certain level of autonomy to complete his research.*"

"I'm well aware of our arrangement, Mr. Madison. However, as the owner of 51% of Troyconics, it is my duty and right to alter that arrangement to establish the status of our present and future discoveries. It has been nearly half a year since you have provided us with a detailed update. So tell me. Are you in control of this situation? Or, as you say in America, are the patients running the asylum?"

There was a moment of near silence as the international phone call crackled with light static. Then Madison's voice came in once more, nervous and shaky. "*Mr. Maychoff, I assure you that I am in complete control of the situation. I have simply noticed that pressure and undue scrutiny tends to hinder Jerry's productivity, not motivate it.*"

"Is that supposed to make me feel better, Mr. Madison?"

There was a sigh on the other end of the phone call. "*I will see to it that a detailed report of Jerry's progress is forthcoming. I'm positive that if there were any major new discoveries, Jerry would have informed me.*"

"Unfortunately," Maychoff told him, "I do not share your confidence in Mr. Jergensen's disclosure tendencies. I know that in many cases, people expect the owners of a business such as Troyconics to be far removed from the day-to-day affairs of the operation. Allow me to be clear: *I am no such owner.* Mr. Madison, do you know how much the electric bill was for your facility six months ago?"

"*Umm, electric bill? Well I don't have that number in front of me at the moment but I assume that it was—*"

"Thirty-six thousand, two hundred sixty-three dollars and twenty-nine cents."

Again a deep silence on the phone, as Maychoff let it hang there for effect. Finally, Madison spoke up. "*Sir, I applaud your attention to detail, but I don't see what that has to do with—*"

"Do you know how much the electric bill for the last three months is?"

"*No sir, but I'm sure that you do.*"

"Are you being coy with me, Mr. Madison?"

"*Not at all, Mr. Maychoff. I only meant to imply that—*"

"Fifty-seven thousand, four hundred eighty-nine dollars!" Maychoff yelled into the phone. "And eleven cents…per *month*. Now. Unless you are completely incompetent, I am sure you know that power usage generally doesn't double when there's nothing to report."

Another brief silence ensues: "*Your point is well taken, Sir.*"

"Good! And Mr. Madison?"

"*Yes sir?*"

"Do not insult my intelligence again. It is my understanding that jobs are still somewhat hard to come by in America these days. Do I make myself clear?"

"*Crystal, Sir.*"

Abruptly ending the call, Maychoff slammed the phone back down in its cradle. A small chip of ivory fell onto the desktop from the phone as it rocked backed and forth until settled. Maychoff slowly stroked his chin in thought and picked up the phone once again.

Troyconics Astrophysics — Ohio, U.S.A.
Earth Date: October 18th, 2014 — 2:43pm EST

Calvin Madison sat at his work desk reviewing department power usage for the last six months. Sighing in anger, he began printing out the documents that Maychoff had been referring to the night before. "What are you doing to me, Jerry?" he muttered, before grabbing the papers, pushing himself out of his chair and leaving the sanctuary of his office.

He walked through the administrative offices in a targeted, mission like pace. Calvin was not happy and his face and walk made no attempt to hide his displeasure. "Good morning Mr. Madison", one lady dared to speak as he continued his infuriated stride through the hallways —and then opened a code door to a secluded staircase down to the lower levels, where the real work of the R&D department took place.

After passing through the security checkpoint, he wound his way through the fluorescent corridors and down the giant elevator to the underground tunnel that led to Jerry's lab. He input his admin override code and stepped back as the giant blast doors rolled slowly to each side.

As he stepped into the cavernous room he spotted Jerry standing by a mainframe tower, talking to someone. It was Ben Levin, Calvin realized. Although a talented young physicist, he was never the less one who didn't have the security clearance to be in Jerry's lab- or so Calvin thought.

"So the Mirror Blades spin and rotate according to this algorithm I developed. It's quite complicated really, every time one of them changes position—"

Calvin simply stood afar off and stared at Jerry, ignoring Ben. The two scientist turn to him, completely caught off guard. The fact that they hadn't heard the blast doors rumbling open meant that whatever they were talking about was extremely engaging. Ben looked around nervously.

"You're not supposed to be down here, Ben," Calvin scolded them. "You don't have proper security clearance."

"I submitted his name for a clearance upgrade already," Jerry covered for his co-worker.

"Did I approve it?"

"Not yet, but—"

"Then he doesn't have proper clearance, Jerry!"

There was an awkward moment of quiet as Calvin's last few words reverberated around them. Finally Ben gave a sheepish nod. "I think, uh… I think I'll head back to my office now."

"Good idea," Calvin told him as he walked past and headed for the gap in the blast doors. Then, Calvin turned back to Jerry, who looked ready to defend himself. "What is this?" he asked, throwing the sheets of expenses towards Jerry.

"What's what?" Jerry asked, picking up a few of the papers and briefly scanning them.

"You tell me. Here I am, doing my best to give you the freedom you ask for and when you discover something big, you don't even have the common decency to tell me! I have to hear about it from one of the owners who rips me a new one right after dinner at my home"

"Calvin I really haven't shared it with anyone, especially the owners."

"And that's the problem!!! Electric bills don't lie. You're up to something. And you can't say you didn't share it with anyone. You shared it with Ben Levin, and he isn't even supposed to be here.""

"Calvin just give me another month and-"

"You really don't get it, do you?" Calvin said. "You're exploiting our friendship and my trust. If I don't get a detailed report from you one week from today… Jerry, I'm shutting you down."

Jerry flinched at the ultimatum. "You can't be serious! You have no idea what I'm on the cusp of—"

"And that's the problem! I have no idea! I'm your boss and I have no idea! Maychoff threatened my job!"

Stunned speechless, Jerry simply gaped at his colleague and cohort. Calvin shook his head and turned away, heading out of the blasts doors. When he was almost there he stopped and turned back. "One week!"

Jerry looked from his boss to his hulking creation. When he heard the doors begin their motorized churning, sliding closed, he picked up a file folder and ran after Calvin. He squeezed through the shrinking gap between them just as Calvin was reaching the elevator doors. When Jerry finally caught up with him, he was repeatedly hitting the call button and muttering to himself.

"Calvin!" Jerry cried. "Look… I was wrong. I didn't mean to exploit you. That was not my intent at all. I would have never knowingly put you in such a compromising position."

"Jerry, we agreed that you would keep me informed of any major developments. It was that agreement and that trust that got you all of this funding and all of this… freedom. You betrayed that trust."

"I know but I am on the cusp of something that can change our nation, … maybe even the world!"

"How so?" Calvin retorts.

"You gave me a week," Jerry said. "Let me compile a detailed report and I'll bring it to you by Friday. I promise I'll explain everything. At least everything that I can figure out by then."

Calvin sighed. "Please tell me why I get the feeling there's a huge 'but' coming?"

"Because I need a few things to make it as detailed as possible."

"Like?"

"A few personnel—Ben included—and some equipment." He held out a document of his own. A list of candidates for a project team and requests for hardware. Calvin took it and scanned the names on the list.

"Since when are fourteen people a few? And this equipment costs hundreds of thousands of dollars. Atmospheric processors, environmental suits… a lot of this stuff will probably be on backorder.

"I've already ordered and paid for the equipment. I used the money in my general R&D fund. Unfortunately, some of the items are on backorder. The shipment's been delayed. I was hoping you could use your pull to expedite the orders."

Calvin gaped at Jerry. "What about the personnel? Dr. Bevilaqua is in Italy and I'm sure she's not just sitting around waiting for you to call."

"Let's just say she owes me a favor," Jerry said, dismissing Calvin's concern. "Besides, this is right up her alley. Trust me, she'll want in."

"And the others?"

"They're all local. Not as high profile, but just as good."

"You've got it all figured out, haven't you, Jerry?"

"Pretty much," he said, unable to stop the excited grin spreading across his face. "At least on this side of the planet."

"What is that supposed to mean?"

"Nothing. Don't worry, it'll be in the report."

The elevator arrived with a loud *ding* that echoed throughout the concrete tunnel. Calvin stepped into the car and turned back to Jerry. He sighed. "Why do I keep letting you do this to me? Alright Jerry! Fine! If you can pull it off and have the full report by Friday, I'll pull all the strings I can to get you your equipment."

"Calvin thanks!" Jerry said excited. "I promise you won't regret this." Jerry steps into the elevator car, hugs Calvin and gives him a kiss on the cheek and then steps out of the car. Calvin gives him a look of concern.

"Uhhh Jerry…. Calvin responds, We are not in Europe and I don't know when was the last time you've seen Candy but perhaps you should go home soon… like now."

Jerry walks away, "I love you man!" as he disappears out of Calvin's sight.

"Lord please don't let me regret this!" Calvin said as the elevator doors close.

Jerry Jergensen's Automobile — Columbus, Ohio, U.S.A.
Earth Date: October 18th, 2014 — 3:41pm EST

"*Hello?*" Ben asked in Jerry's ear. Even turned up to full volume, the car's bluetooth speakerphone was barely audible over the steady percussion of rain beating on the windshield and the screech of his wipers. Jerry could hardly turn the sound up with the buttons on his steering wheel and still focus on the road.

"Ben? It's Jerry."

"*You still alive, Man? Calvin looked* pissed."

"Yeah, well he still is. Somewhat. But that's what I'm calling you about. I need your help."

"*With what?*"

"I'm putting together an exploration team to go through the mirror and I need them by Monday."

"*You been smoking the devil's lettuce, Jerry?*"

"Most of the guys are local. You know a few of them. Benji Sabine, Paula Shuf, Mario Beltose, a few others. I'll email you the rest but I need you to get on the phone starting right now and use the old carrot on a stick with these guys. No details about what we're doing, just enough to bait the hook."

"*Are you sure I'm the one who should be doing this?*"

"Don't worry, you're not doing it all alone. You take care of the little fishes and I'll take care of the big one."

"*Okay. I'll wait for the list. Later.*"

"Thanks, Ben. Later."

A soft *click* told Jerry that the conversation was over and he thumbed the button that activated the voice commands on his phone once again. He took a deep breath as the bluetooth controls came to life. "Call Lilliana Bevilaqua," he told the digital assistant, then waited as he heard Lilliana's phone ringing on the other end.

Barzini Hotel & World Conference Center — Vienna, Italy
Earth Date: October 18th, 2014 — 9:46pm CET

"...So again this proves the concept that we initially started with, that string theory automatically contains gravity. Thank you."

Applause erupts from the hotel's packed conference hall as the audience rises to their feet in ovation. Lilliana Bevilaqua, who's Nobel Prize had turned her into a national treasure soaked up the enthusiasm. The lecture had gone perfectly—no misplaced words, no technical difficulties, just good science and a good audience—and as she stepped away from the podium she had to stop herself from pumping her fist in celebration.

Stepping back stage and out of the audience's sight, she allowed herself to catch her breath. Her beautiful, long black hair fell in tumbling curls down past her shoulders. Soon there was an electronic buzzing sound coming from her clutch. She opened the bag and took out the phone, looking at the screen.

"America?" she muttered, intrigued. Then she thumbed the answer button. "Hello?"

"*Lilliana, my friend,*" Jerry Jergensen started, in his best Don Corleone voice impersonation from The Godfather. "*Are you ready to do me the service?*"

She smiled. "I don't know Don Jergensen," she told him. "Is it legal?"

"*Of course... not!*" Jerry exclaimed.

"Then I'll have to decline." Lilliana jested, playing along with Jerry ruse.

"*Now you owe your Don a service.*"

One of her eyebrows shot up, both intrigued and frustrated by Jerry's vagueness. "Okay, Jerry. Cut the crap. What do you want?"

"*Well I'm happy to hear from you too Lilliana.*"

"Listen, I'm booked for the next two months and for you I might just be booked for the year."

"*I don't believe this. Where is the love?*"

"Goodbye, Jerry," she said, getting ready to hang up.

"*No, wait. Okay, I'll cut the crap. Here it is: How would you like the opportunity to be part of the first team to explore a parallel world."*

Lilliana could feel her mouth drop open. "Oh my goodness you did it. You got the mirror to work."

"Even better. I've already been thru once. Albeit very briefly."

"When do you need me?" she asked, without even a second thought.

"So I take it that you're no longer booked?"

"Oh, I'm booked alright... on the next flight out to Ohio. I'll see you tomorrow afternoon."

"Your Don thanks you," he told her, the bad Italian accent coming back in full force.

"If what you're telling me is true, I should be the one thanking you Jerry. See you tomorrow." She hung up and dropped her phone back into her bag. Tilting her head back against the wall, she smiled thinking about what Jerry had accomplished. If it was anything close to what she imagined, it was going to change the world forever. And now she would be a part of it.

Ben Levin's Home
Earth Date: October 18th, 2014 — 4:02pm EST

Ben was at home on his Mac making phone calls to people on Jerry's list. Next up on the list: Benji Sabine an electrical engineering specialist.

"Hey Benji. Ben Levin here. Hey I'm calling on behalf of Jerry Jergensen."

"What's Jergensen up to now?"

"Hey he's made a pretty big discovery and is putting together a team and was hoping you'd join him. Can't give a lot of detail."

"Really? Don't need detail. I know Jerry. I'm in."

"Great! I'll send you the information in a few moments on email."

"Jergensen's been working on this thing for a awhile. Hopefully he's further along than he was last time. But… I trust him."

"Cool! Thanks Benji." He hung up the call. Ben smiled pleased with himself. Seven down six more to go! Siri call Sara Choremo!"

"Calling Sara Choremo."

Sanctuary of Us — Dabar
Date & Time Unknown

The ThreeofUs could sense and see what was coming. As could the OneofUs, who stood beside him on the balcony terrace once more. In the tranquil quiet of the mostly unpopulated planet, they could hear the wind move thru every blade of the perfectly formed grass. Their Oneness now seeing the need to converse about what was to come.

"So they are coming," ThreeofUs commented dispassionately.

"Yes. Indeed they are," OneofUs confirmed.

"How many do we see?"

"As many as we see."

"Then many?"

"Many."

"And this is just the beginning," the OneofUs and ThreeofUs spoke in perfect unison, "We are ready."

"And what will it be like… this time."

"Like the pilgrims and the native Americans," the ThreeofUs relays.

"Indeed. Yet with a slightly different ending." The ThreeofUs turns and looks toward the point where the portal appeared. The OneofUs stared ahead with no change of expression.

"They have a choice; they can make things… different," ThreeofUs declares as he walks back into the sanctuary.

"Indeed!" the OneofUs agrees returning into the sanctuary as well.

Troyconics Astrophysics — Columbus, Ohio, U.S.A.
Earth Date: October 20th, 2014 — 7:08am EST

The mumbled clamor of different voices filled the modern conference room, bouncing off the glass windows and blank whiteboard, making it seem like the room was filled with fourteen hundred people, instead of just fourteen. They were seated around the long conference table, all of them curious about why they were there and what they would be doing.

Finally, Jerry entered, holding a stack of manila file folders. Handing them over to Ben Levin (who in turn began passing them out to the other team members), Jerry addressed the gathered cluster of fellow scientists and other personnel. "Listen up everybody," he told them, waiting for their individual conversations to die down. When he finally had all of their attention, he began. "I hope you all got a good night's sleep last night and found your accommodations sufficient?"

"Same 'ole Jerry," laughed a buff black man from the back corner of the table. David Atkins was one of the most feared Marine Corps vets and military security contractors in the world. He and Jerry didn't like each other, but Jerry knew he was the best at what he does and Atkins knew Jerry paid well. "No good morning. No nothing. Just straight to it."

"Aw, Atkins, that new sensitivity training at the Corp must be getting you," Jerry quipped right back. "You need a hug before we get started?"

"You can hug my ass after a dump."

"Oh, I see you've joined the 'don't ask, don't tell' crowd too."

Ooooohs and aahhhs rippled through the conference room. Atkins gave Jerry a deprecating sneer. "You know I I don't like yo' ass. I'm only here for the money."

"Well you must be my brother from another mother, cause that's why I'm here too," said Victor Dawkins, a scrawny physics specialist. He had a gift for figuring out mathematical equations in his head that most people needed a black board just to write down. He held his fist out across the table to Atkins and the two men shared a fist-bump, laughing.

"Great, just what we need," a beautifully demure voice said, the words almost musical in the woman's lilting Irish brogue. "Dumb and Dumber."

It was Sara Choremo, who was already reading through the file that Ben had handed her, trying to make sense of the information held within. She was smart, small and testy.

"You say something little lady?" Atkins asked from three seats down. Atkins and Choremo had unfinished business from a previous mission they worked together. The wound still seemed fresh.

Sara stood up, looked at him and grabbed her crotch masculinely. "I got your little lady right here. You don't scare me none. I won't even tangle with you if you cross me again, I'll just kick you in the gonads so hard I'll put your nuts in your chest."

There were more *oohs*, some *ahhs*, and some general laughter from the rest of the team. Atkins laughed along with them. "Sara! You can't still be mad at me for tasering you on that ADAT mission, can you? You were freaking out; I had to do it—"

Sara kicked her chair out of the way and lunged towards Atkins but was restrained by the two people between them. "You piece of—"

"Ladies! Ladies!!!" Jerry yelled, silencing everyone with a single bellow. "Everyone calm down. Now I knew I was taking a risk by putting some of you together. Obviously there's still some bad blood. Now if that means some of you feel like you can't work together, please let us know now. We can't afford any disharmony with what we're about to do."

He looked over them, waiting for a response. Finally, Atkins shrugged. "She mad. I'm good."

"Sara?" Jerry asked.

Still glaring at Atkins, she finally turned, sat down and shrugged. "I'll kill 'im later."

"Good! That's good," said Benji Sabine from the other side of the table. "Postponing a murder is always a good idea."

"If we can move on, ladies and gents?" Jerry said, glancing at his watch, and then at the glass wall, where Calvin was watching the entire farce play out. "What now..." he muttered as he saw Calvin walk over to the door and poke his head into the conference room.

"Uh, Jerry, can I speak with you a moment?"

"Uh, sure. Right now?"

"It will only take a moment."

Jerry turned to the assembled team, slightly frustrated. He had just settled them down and now he was going to have to do it all over again. "Excuse us. We'll be right back."

He followed Calvin out into the hallway, where Calvin turned, suddenly in full boss mode. "Jerry, I was not made aware of the unique makeup of some of the people you've selected. And there appears to be some bad blood from other missions. We can't afford to have that spill over which could cost us or endanger the project or even the team."

"Calvin, these guys are professionals. I wouldn't put them in the field if I wasn't sure they could handle it."

"And are you sure? 100% sure?"

"They'll pull it together for the project."

With a heavy, reluctant sigh, Calvin nodded his approval. "Okay. But at the first sign of trouble I want you to pull the plug on Atkins and Choremo. I think she's serious, ... even if he's not."

"Fair enough," Jerry said, nodding. "Point taken.

Before Calvin could come up with some kind of a retort, Jerry disappeared back into the conference room to address his new team. Calvin follows behind.

"Okay, lets get down to business," he told them. "I called you all here because even though some of us are out for money, most of us are in it for the science and the opportunity to be part of something... well, let's not mince words. Something *big*.

"As some of you may have known, I have been exploring the possibility of alternate or parallel worlds through what I've been calling a Dimensional Mirror. We can get into the technical stuff in more depth later, but essentially the concept is bending and reflecting laser light to create a rift in the fabric of space time, pulling something we believe is far away very close to us. The idea was to find a way to bridge the gap between our world and a theoretical parallel planet."

"Jerry," another team member—Rajin Kadesh—spoke up, "you've been talking about this for years now. What's different now?"

"Simple," Jerry grinned at them. "I got it to work."

Jerry begins handing out project mission folders. There is a murmur of excitement and skepticism. The team members begin flipping through their files with eager curiosity.

"Now," Jerry continued, "each one of you will pair up with their Troyconics counterpart to prep and read up on your assignment. We leave first thing tomorrow morning."

"Wow! One-day turnaround..." said Morris Wiemer.

"If you didn't already know your job you wouldn't be here, Wiemer," Jerry reassured him. "Rajin? Ian? I hope you guys are still good at pulling all nighters. Come morning I want you to know the equipment better than the people who helped build it."

"We'll get it done," Ian told him. "A lot of this stuff looks similar to the design we worked on together years ago. All that said though, Jerry, it might not be a bad idea to send a probe in first."

"I'll second that notion," Rajin said, an ounce of concern registering in his voice.

"I appreciate your caution gentlemen, but it's been over six months since I created this device. I've had plenty of time to tweak the process, get out the bugs and learn as much as I can. Plus I've already gone thru once. I don't believe there's any threat."

"Well then let's talk money," Rajin switches the subject back to the design conversation "Where's my check for my design work?" Jerry walks up to Rajin smugly, "Rajin, you said it would never work and that you wanted nothing to do with it. And then you walked out. You know what that means? No check for you!"

Rajin smolders slightly but concedes. Ian smiles. "I always had faith in you Jerry. It's still not too late to send in a probe."

"Let's get to work!" Jerry tells them both, slightly overconfident.

Troyconics Astrophysics — Columbus, Ohio, U.S.A.
Earth Date: October 21st, 2014 — 6:53am EST

There was a quiet, nervous atmosphere in Jerry's laboratory as the team suited up. Newly designed environmental suits slid up over twelve different pairs of legs. . Twenty-four arms slid into the sleeves, then twisted the metal cuffs of their protective gloves to seal in the air. Air gauges were checked, helmets locked on and comm links were tested.

It seemed they were ready.

Jerry stood at the central control console, the expression behind his helmet's visor one of sheer excitement. "Okay. Everyone suited up?"

He listened as the team called off their confirmations in order. Jerry nodded, approving their pre-entry checklist. "Alright then. Powering up Mirror Blades now."

He began inputting commands on the monitors, just as he had the first time he had shown Ben the device. Soon the mechanical *whir* of the motors filled the humongous room. Wind, created by the spinning panels of glass, buffeted the team as they watched in wonder.

"Mirror Blades have reached perfect synchronization," Ian announced, reading from a screen off to the side.

"We're good on power sources," Rajin confirmed. "Motors are running smoothly."

Jerry nodded. "Activating laser light!" He punched a button and the familiar dazzling effect of the red laser splashed across everything in sight. The bubbling blob of light grew within the blur of mirrors until it flattened and opened up to reveal the new world sitting there, just a few steps away.

The team watched in stunned silence. It was Morris Wiemer who finally said what everybody was thinking. "Man, this is weird," he said.

Stepping into line with the rest of them, Jerry took a deep breath. "Alright, people. Let's go." Then, leading the team, he began walking toward the open portal. His team followed just a few feet behind, twelve men and women, undeterred by the overwhelming possibilities that lay on the other side of the Dimensional Mirror. They stepped in, Weimer the only one pausing. He cocked his gun and stepped in.

Dimensional Mirror Insertion Point — Dabar
Date & Time Unknown

There was a slight *crackle* in the stillness of the valley. Then, all at once, a slice of light and air split open, and the portal that ThreeofUs had watched from the terrace of The Us sanctuary reopened. Jerry and his companionswere walking toward it. They stepped through the portal, some curious, some scared but all excited.

"Alright, everyone. Stick together. Mario, can you give me an atmospheric reading?" The team then began their trek further into the interior of the new planet.

"If I didn't know better I'd think we never left Ohio," Mario Beltose informed the team through his comm link. He was staring down at a tablet that synced with the sensors on his and the team member's environmental suits. The sensors streamed data in real time.

"Good," Jerry said. "What about you, Raddick? What do you have?"

Cynthia Raddick was facing the tree line, staring at data read-outs on her own device. "I got woods. Woods for miles."

The team curiously proceeded towards the tree line about one hundred feet from the portal.

"Is it me, or are the trees an unnaturally rich green?" Lilliana asked as they approached a nearby tree.

"Copy that," Sara said looking closely at a nearby tree, after realizing that nobody could see her nod through her helmet. "She's right. They're perfect, even the leaf shapes. It's almost like somebody photo shopped the whole forest. Even the grass is flawless, look."

Jerry bent over and looked down at the grass at his feet. They were indeed perfect, each blade sticking straight up. None of them were bent. None of them were any taller or shorter than any other. It was as if someone had hired landscapers to tend to the entire valley just before the team arrived. "Morris? What do you got?"

"I'm not picking up anything," Morris shrugged. No life signatures except for us. Not even any animals. That's weird. Woods, but no animals?"

"That is peculiar," Bryan Johnson's voice came through all of their comm links.

"What's this?" Lilliana asked, as Jerry turned to see her approaching what at first glance looked like a small security camera box.

"It's a little metal box," she went on to say, as she got closer to it. The device was attached to the tree, its screen black and dormant.

"Looks man made," Atkins observed, joining Lilliana at the strange object. "Jerry, I thought you said we were going to some parallel planet. We ain't went nowhere except to the damn country."

Beep! The metal box suddenly came to life, a *6* appearing on each of its sides, wrought in red LED dashes. Atkins took a quick step back, and took aim at the box with his firearm. "Okay, Jergensen!" he yelled. "What the hell is that thing?"

"I don't know but don't shoot it!" Jerry told him, his heart suddenly beating like a drum in the echoing confinement of his suit. "We've activated something somehow."

"This is not good," Morris said over the comm links in a distant croak. He was shaking his head. "This is not good, I can feel it…"

The metal box was now counting down. *5… 4…*

"Does anybody else smell smoke?" Lilliana's silky Italian accent asked.

Everyone looked around, scanning the horizon and the tree line for any sign of fire. There was no visual sign of any kind of burning, and yet—

"It looks like we've got a huge fire coming this way," Mario Beltose warned as he stared at the screen of his atmospheric comp-scanner. "And it's moving fast."

"I don't see no fire," Atkins said.

"I'm seeing it too," Morris said, his voice trembling in all of their ears. "I'm getting strong heat signatures building to the East."

They all turned in that direction, following Morris' finger. The fire moved towards them swiftly. Bryan Johnson stared at the fire, "We're dead!" he exclaimed.

"See, we're not even here one day and we're already dead." Atkins exclaimed as he grabbed Lilliana's hand and they both begin to run.

The box reads *2*.

Jerry stared at the approaching flames, paralyzed with fear for half a second. Then, as he felt the intense heat buffeting him through his suit, he snapped out of it and took charge of his team.

"Everybody back to the insertion point!" he screamed over the roaring of the fire. "NOW!" They all begin running but the fire moved with a quick deliberateness towards them.

The box reads *0*.

As soon as the countdown reached its end Atkins drops dead still firmly holding Lillianna's hand, his grip dragging her to the ground as he fell and rolled on top of her. Johnson dropped dead as well. The flames surrounded them quickly. Smoke swirled and howled as the blazing inferno seemed to charge toward the remaining team members.

"Let me go! I can't get away!" Lilliana gasped prying at Atkins dead hand wrapped around her small wrist. The fire enveloped them both. She screams in agony.

Rajin screams, "Lilly!!!" as the fire moves quickly towards him. He declares, "I will not die here!" The flames mysteriously seem to go completely around him. He gets up and runs again towards the insertion point.

Together they turned back, running for the gaping portal that would take them back home. Back to their families, back to safety. The bursting thunderclaps of tree trunks combusting in the intense heat provided an almost operatic soundtrack to their mad dash for safety. Only ten yards from the insertion point, the flames began to lick at the heels of the slowest team members.

Jerry reached the portal first and stopped, ushering his team through before going through himself. As he waved Morris and Sara through, he watched the ferocious forest fire chase down the others. Rajin was the farthest from the portal, and the fire had already moved in front of him.

Rajin looked forward as Jerry yelled, halfway between him and the insertion point. In a split second, the Indian man made his decision and continued sprinting towards the portal. Heading straight for the fire, which was ahead of him he moved to step through it.

Then the tidal wave of fire crashed over him... again, and yet he was not consumed or burned. Instead, the flames seemed to pour around him like a river flowing around a boulder. Jerry watched with amazement as the inferno simply ignored Rajin, passing over and around him until he was standing safely in the open valley once more.

But still the fire continued forward. He decided not to wait and see if he would be as fortunate as Rajin who was still running towards him. He turned back to the portal and stepped through, leaping back into—

Troyconics Astrophysics — Columbus, Ohio, U.S.A.
Earth Date: October 21st, 2014 — 7:04am EST

Jerry could feel the heat of the flames kiss his heels as he dove headfirst out of the Dimensional Mirror and landed with a crashing clang on the mesh gangplank that had been erected in front of it. He rolled, hitting himself over and over, in frantic fear that the sheer heat of the fire may have ignited his suit. When he was sure he wasn't going to be burned alive, he groped for the latches on his helmet and removed it, gasping for air.

"Shut it down!" Morris yelled from somewhere in the lab.

"No!" Jerry called out. "No! Leave it open!"

"Are you kidding?" Mario hollered from behind him—though Jerry was too disoriented from his rolling dive to tell exactly where. "You have to shut it down before the fire comes through and blows this whole place to kingdom come!"

"Don't touch those controls!" he reiterated, getting shakily to his feet. "Rajin's still back there. Leave it open."

Sara came over to him and helped him down the rest of the gangplank. "Jerry," she said to him softly. "If Rajin's still back there… There's just no way—"

"He's alive," Jerry told her before repeating it for the rest of the team. "He's still alive, I saw it. The fire just… It just went around him."

"Jerry, that's impossible," Ben said as he stripped out of his suit, exhausted and traumatized.

Rajin stumbled in through the portal as flames licked the portal entrance as well. Every head in the room turned toward it as Rajin stumbled down the gangplank. "Shut it down!!" Jerry yelled. Rajin moved with tired exasperation, but his suit showed no scorch marks of any kind. He was very much alive.

The flames still licked the portal threatening to destroy the lab. "Shut it down, Benji," he demanded.

He tapped in some commands and the portal crackled, as the flames rescinded then disappeared as the Mirror Blades slowed to a whirring stop.

Jerry looked around at the team. There were only nine of them now. He walked up to Rajin, who despite the cheering crowd around him, looked ill. "You couldn't have done anything, Rajin," he said, putting a hand on his shoulder. "She was too far away, there was no chance. "Jerry!" a voice cried, and he turned to see Calvin walking into the lab. "What happened? Where are Doctor Bevilaqua and Atkins? Where's Johnson?"

"There was a huge fire," Jerry tried to explain. "It... It came out of nowhere and Johnson... Atkins and Lilliana were caught in the rear..."

"What do you mean?"

Jerry frowned, looking over his shoulder at his grieving team. Not wanting to discuss it within earshot of them, he grabbed Calvin's shoulder and brought him over to one of the mainframe towers, where they could have some privacy. "I mean they didn't make it, Calvin. They're dead."

All at once the color drained from Calvin's face. For a moment it looked like he might actually faint and Jerry got ready to catch him.

"This is all your fault Jergensen!" Ian called out, walking over to them with fire in his eyes and steam almost pouring from his flared nostrils. "This could have been avoided! We told you to send in a probe first!"

Jerry was getting ready to apologize—to admit that yes, it was his fault and he would never be able to have a good night's sleep ever again—when Ian threw a right hook. His fist connected with Jerry's jaw with a heavy *thud* and pain exploded across the left side of his face as he hit the floor.

"Ian, stand down!" Calvin yelled, finally regaining control of himself. The team followed them over to the mainframe tower and as Jerry massaged his jaw he realized that this was something they were going to have to deal with together. He turned to Calvin. "No, he's right," he said. "I deserve it. It was my fault."

Calvin shook his head, overwhelmed by the entire situation. He took a deep breath and then began giving out orders to the discombobulated group. "Ben, lock down all the servers and access to this data. I don't want anybody going in or out of here without my say-so until we contain this and figure out what happened. Not a word of this leaves this room. Everyone report to medical and get checked out, then report to the dormitory. I want everyone gathered for a full debriefing after medical. Benji, I'm going to need all of the footage you have from this little excursion."

"I'm on it," Benji nodded.

"You alright?" Calvin asked, turning to Jerry and putting an arm on his shoulder.

"I'm fine," Jerry said hollowly.

"I shouldn't have to tell you this, but all travel is suspended without my express authorization. Jerry, this could get ugly."

"Calvin… it already has."

Chapter 3

Standing at their usual post at the balcony railing, ThreeofUs and OneofUs watched the fire the humans had caused burn its way across the valley, leaving a trail of smoldering ground and ashen trees. It had gone past the humans' insertion point some moments ago, and yet it continued to burn, devouring the trees and grass as it swept across the otherwise healthy land.

"We wish we could have interfered?" the two of them spoke together, their mystical voices somehow making the words a question and a statement at the same time.

"Yes, but such an action, especially at this time, would only lead to more death," OneofUs mused.

"Agreed," ThreeofUs nodded.

"They will learn," said OneofUs. "As we did."

"Indeed."

Together they looked out on the charred, scorched land. The fire still raged. "Would you like Us to take care of it?" OneofUs asked.

"No," ThreeofUs replied. "We will do it."

OneofUs gave the other being a meaningful nod and left him, wandering back into the castle-like sanctuary. The ThreeofUs continued to stare out at the wrecked valley and the fire that still burned beyond it, sending a billowing column of black smoke high into the atmosphere.

ThreeofUs closed his eyes, relishing in the blank darkness, and began to speak softly under his breath. *"Fire, cease! Land, grass and trees, be as you were."* As the mumbled words fell from his mouth and floated away on the breeze, the destroyed land below became green once more. The trees that had been gnawed to nothing by the hungry teeth of the flames were regenerated, blooming to life once again as leaves and blades of grass sprouted into fresh existence.

By the time ThreeofUs was done speaking, the valley had been returned to its green lushness and perfection. ThreeofUs opened his eyes and looked out at the restoration. Satisfied, the ThreeofUs turned away from the terrace and entered the sanctuary.

Troyconics Astrophysics — Columbus, Ohio, U.S.A.
Earth Date: October 21st, 2014 — 3:22pm EST

Locked down in the dormitory, it hadn't taken the team long to go a little stir crazy. As they settled down, some talking, some napping, some just passing the time until they could leave and go back to their loved ones. Meanwhile, Calvin had asked to speak to Jerry privately. The two of them met in the lab away from the remaining team.

"Calvin," Jerry said, shaking his head. He hadn't eaten anything since going through the portal, but he didn't care. All he could think of was Lilliana, Johnson and Atkins. "Calvin, I—"

"Jerry, stop it," Calvin ordered. "It's not your fault. If anything it's mine. I should have taken more precautions before letting you go through there. I shouldn't have given you so much free reign without making sure we did our due diligence."

"We've got to go back. You know that right?"

"Jerry! People are dead! Dead! Do you realize what that means? Especially the death of Lilliana? She's one of Italy's foremost research scientists. They're going to demand an explanation."

"I know."

"Do you? Then perhaps maybe you have a suggestion in regards to what I'm supposed to tell the Italian government. Let's see, how about Lilliana went through an inter-dimensional transport device that's not supposed to exist and visited another world. While there, she was burned alive by an alien forest fire.' He shook his head. "You know, I just don't see that making for a good press conference.

"And that's just the Italian government, Jerry," Calvin went on. "I don't even want to think about how the United States is going to respond. This has federal implications written all over it."

"I don't understand why we even have to involve them," Jerry complained.

"Do you think the Italians are going to help us sweep this under the rug? There are going to be inquiries and investigations. I should also tell you, there very well may be legal precedent for them to press charges. Against you, against me, against the whole company."

"What do you mean?" Jerry asked, suddenly terrified. The idea of going to jail after discovering such an important piece of technology was the stuff of nightmares.

Calvin explained, "There's a clause in the company's federal funding contract, which states that any significant scientific discovery automatically gives the United States first right of refusal and the opportunity to buy the company." He looked around at the strangely angled mirrors that towered over them. "I think they're probably going to classify this as significant."

Jerry sighed. "Is there anything else?"

Now it was Calvin's turn to look terrified. He swallowed. "Oh that's all just the tip of the iceberg," he said. "I still haven't called Maychoff."

Maychoff Residence — Corleone, Italy
Earth Date: October 22nd, 2014 — 1:46am CET

The shrill ring of Bernard Maychoff's cell phone pierced the silent darkness of his bedroom and startled him awake. Grumbling with anger, he groped for his nightstand until he felt the vibrating device and picked it up. "What is it?" he growled into the microphone after thumbing the answer button.

"*Mr. Maychoff, I'm sorry to wake you,*" the voice of his assistant said softly. "*I have a Calvin Madison from Troyconics on line three. He says it's urgent.*"

"It had better be," he said, sitting up in his bed as he rubbed the sleep from his eyes. "Put him through."

There was a series of clicks on the other end. Then: "*Mr. Maychoff, I have Calvin Madison on the line for you.*" And then his assistant was gone, and Maychoff could hear Madison breathing.

"Do you have any idea what time it is here?" he asked sternly.

"*Yes, sir, I'm sorry. I wouldn't have called if it wasn't extremely important.*"

"I get the sense that you are not about to deliver good news?"

"*Well,*" Madison said nervously. "*Part of it is good at least... But you're right. It's mostly bad.*"

"I'm listening."

"*Well sir, I'm happy to report that we have indeed made major progress with Jerry's work. It's really quite amazing what he's discovered. I'm sure you and the other shareholders will be thrilled.*"

"I'll be the judge of the that," Maychoff told him, becoming impatient. "Get on with it."

"*Let me just forewarn you that what I'm about to say might sound like a fairy tale. You're going to think I'm making it up. However, I assure you that this is real. Jerry has invented what he is calling a Dimensional Mirror. And... Well, this device has allowed him to not only discover, but cross over to a parallel world, much like Earth.*"

Maychoff simply sat there in the darkness, his phone pressed to his ear.

"*Mr. Maychoff, are you still there?*"

"Go on," he said, seemingly unfazed. "I'm still waiting for the bad news."

"Well sir, with my approval, Jerry assembled a team of field researchers and other personnel and actually went through the mirror to the other planet. While they were there, there was a freak accident and…. well, quite frankly, three of the team members failed to return. They're dead."

"Is there more?"

"Yes. One of the scientists that died is—was—a prominent researcher from your country. Lilliana Bevilaqua."

Maychoff remained inhumanly quiet. "Mr. Madison, my plane will be arriving at seven o'clock your time tomorrow morning. Be sure you are there to greet me."

"Certainly sir. 7:00 am. I'll be there. Is there anything else?

"That's all." And with that, he hung up the call, and began dialing his assistant. If he were going to make it to Ohio as soon as he had told Madison, he would have to leave right away.

Maychoff's Private Jet — Over The Atlantic Ocean
Earth Date: October 22nd, 2014 — 3:49am EST

As the Falcon 900's jet engines roared outside, carrying the plane as fast as it would go toward the United States and the Troyconics Astrophysics facilities, far below them, the churning surface of the ocean looked like some kind of abstract oil painting, thick with texture.

Inside, Maychoff's phone was still glued to his ear. Madison's phone call had set off a chain reaction of further conversations. This was simply another bullet point on Maychoff's call list, but an important one. He listened as the other line rang in his ear. On the third ring, a woman answered the phone. "*Britair Global, how may I help you?*"

"I need to talk to John Hankins," Maychoff informed her.

"*May I ask who's calling?*"

"This is Bernard Maychoff."

"*One moment, Mr. Maychoff,*" the operator told him. "*I'll connect you.*"

Again there was an awkward pause followed by a few clicks and, finally, the chipper voice of an Englishman. "*Bernard! This is a surprise. To what do I owe the honor?*"

"John," Maychoff grinned through the phone. "How are you my friend?"

"*Doing just swell now! Every time you call me 'friend' that means I'm about to make quite a bit of money. So tell me. What can I do for you?*"

"I need you to look into what it would take to buy up all the outstanding shares of Troyconics Astrophysics."

"*All of them?*"

"The entire remaining forty-nine percent."

"*Swinging for the fences, eh? I'll look into it. Let me call you back. Cheers.*"

"I'm looking forward to it," Maychoff said and hung up. As he dialed in the phone number of the next person on his call list, the jet continued its thunderous arc toward Ohio.

Troyconics Astrophysics — Columbus, Ohio, U.S.A.
Earth Date: October 22nd, 2014 — 5:21am EST

The survivors of the first trip through the Dimensional Mirror were asleep in the dormitory. It had taken most of them hours to finally fall asleep— and many had woken at some point during the night—but now they were all quiet.

All except Jerry. In the lab with everything shutdown, it appeared dark and dank, the only source of illumination was the glowing screen of the large desktop he was staring at. Large over-ear headphones ran from the side of the computer and clamped to the sides of his head as he poured over the footage from the mission.

In the footage, which was taken from the camera embedded in Morris' helmet, Jerry reviewed Atkins dropping his hand to his weapon, wary of the metal box which had just started its countdown from *6*.

Atkins yelled in Jerry's ears through the headphones. "*We're not even here one day and we're already dead!*"

He reviewed the footage until the box reached *0* and the fire erupted from the woods. But it wasn't the flames that killed Atkins or Johnson, Jerry now saw. They dropped to the ground as soon as the box finished its countdown. It seem as if they had just... died.

He tapped a button on the keyboard and scrubbed backward through the video to just before the countdown began. "*—Thought you said we were going to some parallel planet. We ain't went nowhere except to the damn country.*"

Then the box activated with a *beep* and the countdown played out again on the desktop screen for perhaps the third dozenth time. Jerry scrubbed back again: "*—the damn country.*"

Jerry shook his head. "This can't be!" A lingering bit of uncertainty still clouded his mind even as he knew what he was thinking was right. He simply did not want to believe it. It couldn't be. It wasn't possible. Suddenly a voice full of anxiety broke his concentration.

"It's started," a muffled voice told Jerry as he looked up, startled. Calvin stood over him, the pale light of the computer throwing faint blue shadows across one side of his face.

Jerry paused the video and slid the headphones down around his neck. "What are you doing up? What do you mean?" he asked.

"Maychoff will be here first thing in the morning, which is a couple of hours from now." Calvin told him.

"Maychoff's coming here?"

"That's what I just said, isn't it?"

"Did you tell him what happened?" Jerry wanted to know.

"Yes," Calvin murmured through the dark.

"And? What did he say?"

"He said he'll be here in the morning, and that's all he said. I'm as clueless as you."

"So that's it?"

"That's it."

Jerry frowned, looking back at the computer; the frame paused on a shot of Atkins falling down dead. "Well let me know what he says."

He started to slide the headphones back over his ears, but Calvin reached out and took them from him. "Oh no! You're not getting off that easy. He's landing at 7:00 am and you're coming with me to greet him."

"Me? You're the Administrative Director. I'm just the worker bee."

"The worker bee who caused an international—and perhaps inter-dimensional—incident because of the project that he insisted on being solely responsible for."

"Maychoff is management. I don't work with management."

"You do now," Calvin told him. "I'm not asking you. You're coming."

Jerry thought about it, then seeing an avenue for opportunity he brightened up. "Okay," he said. "But under one condition."

"Do you really think you're in a position to be making demands?" Calvin asked.

"I just want to call Candy. I won't tell her anything, I just want to let her know that I'm okay and not to worry."

Despite the poor visibility with only the desktop light, Jerry could see the expression on Calvin's face perfectly. He had seen Calvin make that face for decades. It was the sighing smirk he gave when he was about to give in. "Fine. You can make one phone call. And don't tell the others, or they're all going to want to start calling and emailing and texting and tweeting. This is just the once."

Calvin held out a cell phone to him and Jerry took it. "Thanks."

Jergensen Residence — Columbus, Ohio, U.S.A.
Earth Date: October 22nd, 2014 — 5:31am

The light of dawn was just beginning to peak through the venetian blinds that blocked Jerry and Candy's bedroom window, when all of a sudden Candy's phone began to chirp next to her head. Not even bothering to get up, Candy merely reached out, grabbing blindly for the device. "Hello?" she mumbled groggily into the microphone.

"*Hi honey, it's me.*" Jerry said.

Candy lifted her head and peered over to her left where she saw Jerry's empty side of the bed. He had never come home. This wasn't cause for alarm (Jerry often pulled all-nighters at the lab), but there was something in her husband's voice that seemed worrisome. "Hey," she said softly. "Everything okay?"

"*Yeah, everything's fine. We just... We ran into some issues here at work. I'm going to be stuck in meetings all day and then I'm probably going to have to file a mountain of paperwork. Anyway, as you as you can already see, I'm not going to be able to come home tonight either. I'm sorry.*"

"It's okay," Candy said, sitting up under the covers and leaning against their headboard. "Are you sure everything is alright?"

"*It will be,*" he assured her. "*Just... will you do me a favor? Can you pray for me?*"

"Sure, honey. I always pray for you. I love you."

"*I love you too. We'll talk soon.*"

He hung up and Candy was left alone in the bedroom, the mattress seemed huge with only one person on it. Sighing, she leaned back over to her nightstand and clicked on the bedside lamp there. Then she took the Bible that was laying on the nightstand and began to read.

Troyconics Astrophysics — Columbus, Ohio U.S.A.
Earth Date: October 22nd, 2014 — 6:49am EST

Bnk! Bnk! Bnk! The bounce of a tennis ball hitting a concrete wall echoed through the dormitory as Dawkins threw it and caught it again. Threw it and caught it. *Bnk! Bnk! Bnk!*

"Can you stop that?" Sara asked, still halfway asleep in her bunk, an edge of irritation creeping into her voice.

Dawkins caught the ball and held onto it. "Why?" he asked, and then threw it again.

Bnk!

"Cause it's annoying and it's six o'clock in the morning! You're so rude!"

Dawkins caught it and turned to look at her. He stared at her for a moment, then shrugged and threw it.

Bnk!

"Alright, that's it, I'm tired of this!" Sara started saying, jumping down from her bunk, her nostrils flaring like mad.

But before anything happened, as the ball fell back down, Rajin quickly stepped in, reached out and snatched it from the air, taking it away from Dawkins. "Enough, we're all restless and frustrated!" he told both of them. Sara paused still agitated.

"How long can they keep us here?" Mario complained. He was lying on his own bunk, staring up at the bed above him.

"I'm sure it's just a formality. They're simply trying to contain the situation," Morris offered, sitting on the hard floor, in the middle of a game of solitaire.

"I don't care what they are trying to contain," Benji chimed in as he fiddled thru today's newspaper. "I gotta get out of here."

"Benji's right," Ian said. "They can't hold us against our will like this."

"What do you think is going on?" Morris asked.

"Isn't that obvious?" Ian laughed. "They're getting their cover stories ready. They have to make sure everybody is on the same page and telling the same lies.

"But Lilliana is their biggest problem," he added. "That's potentially an international incident. There's no cover story big enough to cover that up. Lilly was well known all over her country."

"All over the world," Rajin said, staring at the ground. An eerie silence came over the group as the conversation turned back to their dead colleague.

Johnson Air F.B.O, Columbus, Ohio, U.S.A.
Earth Date: October 22nd, 2014 — 7:00am

At exactly seven o'clock, as promised, Bernard Maychoff's Falcon 900B touched down, it's tires screeching against the tarmac. Calvin and Jerry stood together just outside the door of a small hangar, watching the three engine beast roll to a stop and then taxi over towards them. They were at a fixed-base operating airstrip just outside of Columbus.

The jet finally came to a complete stop and shut down. The door unfolded and out stepped a distinguished looking man with salt and pepper hair and a fit build that betrayed his age. Bernard Maychoff took each step with confidence and walked straight over to Calvin.

"Mr. Madison," he said, still seeming to have an attitude from their previous conversation.

Calvin maintained eye contact. "Mr. Maychoff! It's good to finally meet you in person, sir. Though the circumstances are unfortunate."

Maychoff grunted and turned his attention towards Jerry. "And who might you be? I was not aware we would be having guest." Calvin spoke up nervously, "This is Jerry Jergensen, the scientist who made the discovery." Maychoff's eyebrows rise, "Ahh the one who created these unfortunate circumstances, isn't that right, Mr. Jergensen?"

"What happened was a tragedy. I lost good people...friends." Jerry said. "But I stand by my decision to go through the Mirror."

"A man who stands by his decisions, even when they appear to clearly be wrong." Maychoff smirked slightly.

"Why don't we head over to the facility?" Calvin interjected, always the peacemaker. "It will be easier to discuss all this there. Jerry, why don't you bring the car around and we can—"

"I've arranged for my own transportation while I'm in town," Maychoff interrupted. He pointed across the tarmac where a shining Bentley equipped with a driver was just then pulling around the corner of another hangar. It turned and headed straight for them. "We can all ride and talk together in my vehicle."

A few moments later, the three of them were sitting in the spacious interior of the gorgeous, luxury vehicle.

Though Jerry and Calvin could barely get over the lavish styling of the automobile, Maychoff seemed not to notice at all. He simply got down to business.

"So," he said, "We have three dead people. One an internationally famous research scientist from my country, I might add."

"Like Jerry said," Calvin nodded solemnly. "what happened was a tragedy."

"Yes, well, I know this may appear cold to you, but I'll wait for the details of what happened in your report. What I am more interested in is this new discovery—this Dimensional Mirror—and acquiring the remaining forty-nine percent of Troyconics Astrophysics. I'm assuming, Calvin, that you were at least smart enough to have everyone on the team sign a release and indemnity agreements that relieve us of any legal responsibility?"

"Of course," Calvin said. "I have all the necessary forms."

"Good. I'll handle Ms. Bevilaqua's unfortunate death with the proper authorities and statesmen. I have a few people in the interior ministry who know how to handle situations like this. Between them they should be able to keep any ruckus to a minimum."

Then Maychoff turned in his seat and stared into Jerry's eyes. "And now we come to you, Mr. Jergensen."

"Sir?" Jerry said, raising an eyebrow. Ever rebellious, Jerry refused to be unsettled by the rich man's gruff demeanor.

"Although I am elated and excited by the prospects of the discovery Mr. Madison described to me over the phone, I also feel that you are partly to blame for this fiasco."

"I understand."

"Really?" Maychoff asked. "Do you truly? I don't believe you do. If you understood then we would have known of this device weeks ago—months ago—and someone might have been able to rein in your exuberance.

"What do you mean?"

"What I mean is that had you reported your progress to Calvin or another superior, we could have evaluated the risk of sending fourteen people through some portal and possibly made, a *safer* assessment.

"I believe the possibilities that your discovery presented blinded you to the possibilities of what might have gone wrong. What *did* go wrong."

"Sir, with all due respect—"

"Mr. Jergensen," Maychoff cut him off, "before you say another word, allow me to warn you. It's been my experience that when a person prefaces a statement with 'with all due respect', it likely means that they are about to disrespect me. So I warn you: *tread lightly.*"

"Sir…" Jerry started again, choosing his words carefully, "the nature of this discovery and what happened is beyond anything anyone has ever discovered. I've reviewed the tapes and I believe that there are dynamics and principles involved that are not in our case normal. Every possible precaution was taken and ran through Calvin—"

"So you're saying this was Mr. Madison's fault?" Maychoff asked, furrowing his brow.

"Not at all," Jerry clarified. "I'm simply saying that the team we assembled covered every possible contingency. Scientific, military, everything that we knew we could face."

"And yet something went horribly wrong. Which means…"

"That there were things we didn't know, sir."

Maychoff regarded Jerry with judging eyes, trying to determine what exactly he meant by it. Before he could reach a verdict, however, the conversation was interrupted by Maychoff's cell phone. "Excuse me," he said, holding up a finger before answering the phone. "Yes?"

"*Bernie, old boy,*" John Hankins said from London. "*You're going to love me. I got the remaining shareholders to agree to sell their shares.*"

"Good. Anything else?"

"*Well, there is a clause in the paperwork concerning the U.S. Government—more than likely the Department of Defense—having first right of refusal. In the event that a major or significant scientific breakthrough is made, they have the option to buy out the entire company at fair market value.*

I haven't heard anything about any discoveries being made, and I don't want to hear about any before I get my commission so: what do you say? Is the deal a go?"

Maychoff pondered the question for only a second, sparing a glance at Jerry in the seat across from him. "Indeed."

"So no problems on the discovery issue?"

"I will handle any issues," he said.

"Excellent. I'll have all the paperwork sent over to you by tomorrow afternoon. By the way, the price point for the remaining forty-nine percent of Troyconics is valued at one hundred and eighty-one million dollars."

"I'll have the funds wired to you after the paperwork and transfer sale is complete."

"Bernard, that could take a while, and I had to pull a lot of strings to get this going."

"Okay, then I'll have $362,000 wired to your account by morning. Is that enough to get the ball rolling?"

"As always, it's a pleasure doing business with you."

"And you, John," Maychoff said, hanging up the phone. Without looking at Jerry and simply staring out of his window, he addressed Jerry once again. "Mr. Jergensen, your arrogance offends me, but this phone call has put me in a celebratory mood. So, instead of firing you…"

He paused letting the looming threat hang in the air between them as he continued staring out of the window. Although Jerry's face remained arrogantly confident, inside he was deeply concerned and slightly embarrassed.

"You and Mr. Madison can join me for breakfast, and you can fill me in on the details of your project."

Both Jerry and Calvin sighed silently with relief. "Sure," Jerry said, trying to relax a little bit. "That sounds great."

Quilsec Astrophysics — London, U.K.
Earth Date: October 22nd, 2014 — 12:37pm GMT

In another part of the world, at the facilities of another one of Bernard Maychoff's many companies, Brian Ridley was in the middle of a late lunch, when his cell phone began to ring. Born in the U.S., Brian had a bit of trouble getting to know his British co-workers. And so, his lunches had become a rather lonely and silent affair, mostly eaten (as was the case today) in his office. While returning from the restroom, his cell phone rang. Stepping into an empty office he closed the door and answered the call.

"Hello?" the young data analyst answered, eager for the company, even if it was only auditory.

"*Mr. Ridley?*" an English accent asked.

"Yes," he said, now a little wary. "Who is this?"

"*My name is John Hankins. I'm calling with a bit of information that I believe your superiors might be interested in.*"

"Go on…" Brian said, grabbing a pen off the desk and a pad out of his pocket, getting ready to take down notes.

"*First of all, I need strict assurances and guarantees that this will, uh… close any debts due to my past indiscretions?*"

"We'll have to see how valuable the information is."

"*Trust me,*" the Brit said, "*this is top shelf.*"

"I'll be the judge of that, Mr. Hankins!"

"*I would appreciate it if you would refrain from using my name on the line.*"

"What do you have?" Brian asked.

"*Bernard Maychoff,*" the voice said. "*He contacted me yesterday morning concerning a buy-out of the remaining shares of Troyconics Astrophysics.*"

"And how does that concern me and my superiors?"

"*I'm not going to do all of your homework for you. Suffice it to say that he already owns a controlling interest in the company, and the remaining forty-nine percent would give him complete ownership.*"

Brian took a seat on the desk, the phone glued to his ear, waiting for

Mr. John Hankins to keep talking. But the line went silent. "I'm still waiting for why we should care."

"There was a buy-out clause in the company's charter that gives the United States Government first right of refusal to purchase the company in the event of a significant scientific discovery or breakthrough. Meanwhile, the price of the remaining shares came out to about one hundred and eighty million *dollars, which Maychoff was all too eager to shell out the cash and get this deal done. Now I am sure that you know as well as I do that no one pays almost two hundred million dollars for yesterday's toys."*

Brian was still unsure. There was nothing concrete here—just educated guessing—but it was enough to pique his interest, and maybe the interest of his boss. Not his boss at Quilsec, of course, but his *real* boss. "I'll get back to you," he told Hankins and hung up the phone.

Not wanting to waste a second, Brian reached in his suit jacket pocket and pulled out another phone—a disposable phone—and dialed a phone number that he had committed to memory. It rang a couple of times and then—

"Bill Tish," a hard-sounding man answered.

"Yes sir, this is Brian Ridley. You asked me to report any developments concerning Bernard Maychoff to you directly."

There was silence on the other end for a second or two. *"Go ahead,"* Director Tish finally said.

"Sir, I have reason to believe that Mr. Maychoff is in the process of buying the remaining shares of Troyconics, which will provide him with complete ownership and authority over the company and all its projects. The cost of the remaining shares was nearly two hundred million dollars. I also have information that leads me to believe that Maychoff's sudden interest may have been prompted by a recent scientific breakthrough at the company."

"Alright, Mr. Ridley. Good work. I want you to compile a detailed report and gather everything you can from your source, then meet me tomorrow for debriefing. I don't trust electronic communication with information of this priority. Can you get here by tomorrow morning?

"I'll make it happen."

"Good work Ridley. I'll see you in my office at 0800 hours."

"Yes, sir." Brian hung up and put the burner phone back in his suit pocket. Excited and eager, he failed to notice the ominously blinking LED light of the smoke detector overhead, which was fastened to the wall just over the door.

Troyconics Astrophysics — Columbus, Ohio, U.S.A.
Earth Date: October 22nd, 2014 — 1:14pm EST

The TV in the research dormitory provided a distraction from the isolation the team had been subjected to since that morning. Along with the lunch of submarine sandwiches and chips that had been brought down to them, the TV had a calming, sedative effect. Now, instead of arguing and worrying about when they would be let out, the team was simply happy to eat their food and watch the screen.

"*This just in,*" a news anchor from UNN was reporting at the moment, "*last week's fire at the Ohio Space Institute claimed the lives of three research team members. Among the deceased is well-known Italian scientist and field researcher, Dr. Lilliana Bevilaqua. The other two team members have been identified as David Atkins and David Johnson both Ohio natives. The three were part of an experiment gone awry that ended in an explosion and massive fire inside the laboratory. Services for Atkins and Johnson will be held sometime next week. Services for Bevilaqua, while the details have yet to be confirmed, will take place in her home country of Italy. In other news, a local zoo is raising money to preserve...*"

Ian accidentally swallowed and nearly choked on a bite of his sandwich, coughing up the little chunk of un-chewed food. He shook his head, affronted. "Are you serious? That's really all they're going to say about it?"

"What?" Mario said in the seat behind him. "I thought it was pretty good. Simple and to the point. Not a whole lot of room for further questions, just cut and dry."

"I got a question," Morris spoke up. "The Ohio Space Institute is a government facility. Why would they agree to let us use *their* program to cover up *our* screw-up?"

"Yeah," Benji echoed the concern. "Something's not right here. Why would they do that?"

"Who cares?" asked Mario, throwing potato chips into his mouth with a nonchalant *crunch*. "They got the story out in the open. Maybe now they'll let us go home."

"Don't bet on it," Morris said, shaking his head.

The Pentagon — Virginia, U.S.A.
Earth Date: October 23rd, 2014 — 9:02am EST

Brian Ridley was exhausted. After getting off the phone with Director Tish, he had worked for the next three hours straight, trying to compile as many details and facts together before submitting his report to the Director. Soon after he was picked up by a driver and immediately whisked away to Heathrow airport, and put on a plane that would take him stateside. Upon arrival he was briefed by Director Tish, and now he found himself being escorted to The Pentagon for the first time in his career.

They wound through the intricate system of hallways and elevators, making their way deeper and deeper into the depths of the government complex. Finally, far below the ground floor on which they had entered the building, they came to what Brian could only imagine was a situation room.

The walls were wallpapered and bare, with the exception of several large flat screen monitors. At the moment the screens were on standby, the only image on them being a graphic of the seal of the Department of Defense. Around a large conference table, several military personnel (including a pair of five star generals) and political power-players in suits sat around it, waiting for them.

"Ladies and Gentlemen," one of the generals, General Maddox—a weathered man with a thick build and hard eyes—said as he stood and gestured to two empty seats. "I think everyone knows C.I.A. Director Tish. And this is one of our covert field operatives, who works through the C.I.A. in conjunction with us on special assignments, Brian Ridley."

The Director reached out and shook the general's hand over the table. "General Maddox. Thank you for pulling all of this together on such short notice, but I believe it will pay off."

"Let's hope so," Maddox grumbled as Tish sat down in the chair opposite him, at the center of the table. Brian took the empty chair to the Director's right and sat down. Even with his extensive experience he couldn't help feeling like a little kid who'd been invited to play ball with the big boys. He just hoped he wouldn't embarrass himself.

"Donna, bring up the dossier," General Maddox said.

Donna manipulated a hand held touch screen panel. The center screen lit up with a photograph of Bernard Maychoff. Maddox narrated the information that appeared on the screen next to the picture.

"Some of you may know this man. His name is Bernard Maychoff. He is an extremely wealthy businessman who resides in Corleone, Italy. Mr. Maychoff has a profound interest in astrophysics companies—particularly when it comes to R&D. He invests or buys promising companies in the start-up or development phase. Until very recently, Mr. Maychoff owned a fifty-one percent stake in Troyconics Astrophysics in Columbus, Ohio."

The photo of Maychoff disappeared and was replaced with a business profile on Troyconics Astrophysics. The people around the table soaked in the information on the screens, which included a photo of the Troyconics facilities, stock charts, valuation and the names of its top executives.

"Recently, Mr. Maychoff executed maneuvers to purchase the remaining forty-nine percent of Troyconics, using a series of shell corporations. I'll let Mr. Ridley take it from here.

"Thank you, General," Brian quipped, all eyes suddenly on him. "As the General stated, Mr. Maychoff recently formulated a deal to purchase the remaining shares of Troyconics."

The faces around the table stared back at him, eyes blank. He winced, realizing how stupid he must have sounded, parroting back something the General had just said. They would think he was an idiot. He tried to recover, thinking back to the report he had compiled. "Troyconics has been in business since 2001. It was originally funded through a loan from the U.S. Government. In light of that funding, the government took an interest in some of the promising projects being worked on there and included a clause in the loan agreement, which gave them first right of refusal to purchase the entire company in the event that a significant scientific breakthrough or discovery were made.

"Right now we're not sure whether Mr. Maychoff is aware of the clause, as he purchased the company after the loan was granted. However, if he's willing to fork over $181 million for the rest of Troyconics when he already owns a controlling interest, we believe that indicates they've stumbled onto something of enormous value over there in Ohio."

"Thank you, Mr. Ridley," the General nodded curtly. "Now, if Maychoff is aware of the clause, this might be the reason he is attempting to execute the deal before the news of whatever they discovered leaks.

However, based on the language of the clause, it doesn't matter. If news of any discovery is not forthcoming, the government has the right to execute the clause retroactively and even use force if necessary."

"Do we have any idea at all as to what they may have invented or discovered?" asked a woman in a naval dress uniform.

"Brian?" The General called on him again.

"At this time we do not. However, we are visiting the Troyconics facilities as we speak, to talk with Calvin Madison. He's the Administrative Director, who oversees the entire facility. We hope that he'll be... cooperative."

"And if not?" a silver-haired man in a suit asked from the head of the table.

"Then we'll have to find out the old fashioned way," the General said with a stoic tone of finality.

"Force?" the Navy woman asked.

"No," General Maddox grinned. "Blackmail.... in a way. Mr. Madison has a close friend over at the Ohio Space Institute. Today the local news reported that three research scientists were killed in an explosion at the O.S.I. labs. Now, our inside sources tell us that there was an explosion and that some people were hurt, but nobody was killed. And to their knowledge, neither of the persons identified in the media story were anywhere near that location. This leads us to believe that the whole story is a desperate attempt at a cover up and that whatever this discovery is; it was responsible for the death of these three people. If our suspicions turn out to be true, and Madison did pull some strings to cover up these deaths, then we'll have plenty of leverage on him and whoever else is involved."

Brian noticed Maddox shared a meaningful glance with the silver-haired man at the head of the table, locking eyes for a moment before breaking away. Meanwhile, the female naval officer continued her line of questioning. "General, is there something you're not telling us? There seems to be a fair amount of heavy hitters at this table for a simple company take-over."

"Major," Maddox said with a snort, "as usual your woman's intuition has sniffed out the other side of this issue. The company is only part of the problem here. There is more to it than just some company. It's come to the attention of the United States that some of Maychoff's companies have had some dealings with several known terrorists. Now, we're not sure whether he's aware of this or not, frankly, with the amount of ignorance on our part and the amount of money Maychoff is willing to part with, we don't believe it's in our best interest to handle this at a lower level.

Better to be over-prepared, I always say. Which is why we're working side by side with the C.I.A. and the N.S.A"

When the General said *C.I.A.* he looked over at the man with silver hair again. When they broke eye contact this time, the man at the head of the table looked over at Brian, catching him staring, and Brian looked away, terrified.

"Well," Director Tish said, "as the General mentioned, we currently have two N.S.A agents on their way to Troyconics to speak with Mr. Madison. Once they interview him, we can decide how we need to proceed.

Troyconics Astrophysics — Columbus, Ohio, U.S.A.
Earth Date: October 23rd, 2014 — 10:28am EST

Calvin sat at his desk, going over Jerry's written report of the tragic trip through the mirror. He had called in some humongous favors over the past forty-eight hours, trying to cover up Lilliana's and the others deaths.

An electronic *ringing* spat suddenly from his office phone and he picked it up, agitated. "What? Hello?"

"*Mr. Madison, there are two gentlemen here at the front gate.*" Calvin glanced at the phone's caller ID display and saw that the call was coming from the extension for security. "*They have N.S.A. credentials and they say they're here to see you, but we're not seeing them on the books.*"

"That's because they're not on the books," Calvin said, suddenly terrified. What did the N.S.A. want with him? What did they know?

"*What do you want us to do, sir?*"

Calvin took a deep breath, thinking. "Send them to my office," he finally said.

"*Yes sir.*"

There was a dry *click* on the other end of the line as the security guard hung up the phone. Calvin dropped the receiver back into the cradle and got up. Figuring he had about five minutes before the agents made their way to his office, he ran down the hall to the men's bathroom. Inside, he splashed water on his face and cleaned himself up as he had begun to sweat profusely. He straightened his shirt collar and brushed his fingers through his hair, making sure he was presentable. Then he sprinted back to his office and sat behind his desk, trying to look as though this were just another day at work.

He beat the N.S.A. agents to his office by about thirty seconds, and stood when they entered the room. The two men appeared to be both in their thirties, and both donned impressive black suits. Escorted in by two security guards, they shook Calvin's hand in turn.

"Mr. Madison," the one on the left said, "I'm Special Agent Sanders. This is Special Agent Brown. We're here on what we believe may be a matter of national security."

"Wow, you guys get right to the good stuff, huh?" Calvin tried to joke, chuckling nervously.

The two agents didn't laugh at all. In fact, they ignored the comment completely. "Sir, there was a recent story in the news," Agent Brown went on. "Maybe you read about it: three people died in an explosion at O.S.I."

"I…" Calvin started, then looked down at his hands. "Yes, I know all about it. It's quite tragic really… If you knew Lilliana… Really a shame."

"Yes, it is," Agent Sanders nodded. "The problem is, Mr. Madison, we're getting conflicting reports regarding the details of the incident."

"Conflicting?"

"Conflicting," Brown said.

"We have reports saying that Lilliana Bevilaqua was never actually at the site of the explosion. We have reports saying she was here at Troyconics."

"What?" Calvin said, trying not to show any sign of doubt in what he was saying. "Certainly not. Ms. Bevilaqua was nowhere near our facility. Why would she be?"

"Mr. Madison, are you aware that lying to a federal agent is a felony?"

"I wasn't, actually," Calvin snapped back, acting insulted. "I haven't had any run-ins with the law."

"Well it is. So, do you care to change your story?"

"Gentlemen, am I under arrest here?"

"No sir. You're not …yet.", Agent Sanders said grimly.

Calvin didn't crumble. "Well, good because I believe it best to relay this information to our higher ups."

"And who here is higher up than you Mr. Madison?" Agent Brown inquired. "Ahh, you mean Mr. Maychoff."

"Are we done here gentlemen?", Calvin asked, eager to end the conversation.

The agents rose from their seats with an overconfident smirk. Agent Brown dropped a business card onto the top of Calvin's desk. "Tell Mr. Maychoff we'll be in touch. Real soon." he said with a grin as the two N.S.A. agents left the room.

Calvin sighed deeply. "Crap!"

Meanwhile, outside the door and out of earshot, Brown and Sanders conferred with one another. "He's lying," Brown said.

"What was your first clue?"

They began walking down the hallway, when they passed two men—it was Jerry and Maychoff—walking in the other direction, back toward Calvin's office. "So actually the mirror acts as a sort of a doorway to…" Jerry said as their paths crossed. Maychoff eyed them both suspiciously.

Once the men were out of sight around the corner again, Brown turned and stopped his partner. "That was him," he whispered.

"Who?" Sanders asked.

"That was Maychoff!"

"I thought he lived in Italy."

"He does, but that was definitely him. Something's going on here. And for him to come all this way, it must be something big…"

As the two N.S.A. agents speculated on the significance of Maychoff's presence, Jerry and Maychoff were knocking on Calvin's door. "What?" yelled a frazzled voice from inside, and Maychoff opened the door.

"Who were those men?" he asked as he let Jerry pass and closed the door behind them.

"Those are two agents from *thee* National Security Administration!" Calvin said, his voice on the brink of turning to hysterical mush. "They were here asking me questions! Interrogating me about the OSI story!"

"What did you tell them?" Maychoff demanded.

"They said they were getting conflicting reports about Lilliana's whereabouts at the time of the O.S.I. explosion. They said that someone told them she was here. I told them she wasn't here."

"Oh, Calvin, why did you say anything?" Jerry groaned. "You should have kept quiet, asked for a lawyer. Don't you know lying to a federal agent is a felony?"

"How was I supposed to know that? I'm not a criminal! Look, we have got to figure out a way to bring this out in the open and get past it."

"No, that's impossible," Maychoff grumbled. "I have already provided the people in Italy with our cover story. We lose all credibility if we change our story now."

"Well that's going to do a lot of good if this thing blows up and the truth shows up on CNN," Calvin shook his head. "What do we do in the meantime?"

"Did they leave you a card?" Maychoff asked. "A way of contacting them?"

Calvin held up the business card, handing it over to Maychoff.

"Do *not* speak of this to anyone," Maychoff said, tucking the card away in his breast pocket. "You two do your jobs. I will deal with the N.S.A."

"What are you going to tell them?" Jerry asked.

"The truth! Meanwhile, there is something I need you both to do."

Chapter 4

"Well, it's about freakin' time!" Sara cried when the dormitory door opened and Jerry and Calvin entered. The rest of the team had all been waiting for them and now they stood, eager for answers and on the verge of mutiny.

"Sara, please," Jerry said. "I thank you all for your patience and understanding. I think you know it was necessary for us to hold you all here. And in light of the most recent developments, I'm sure that we made the right decision, although I know it was uncomfortable."

"Man, my wife is going to be pissed," Morris Weimer complained. "What the hell is going on?"

"Calvin, what the hell?" Benji asked. "I hope you guys are here to tell us that we're free to go.

"Benji, I apologize but I am not," Calvin told him.

"What? Why not? This is bull—"

"Benji look," Calvin cut him off. "This is serious! The N.S.A. is involved now."

"The N.S.A like government N.S.A?" Benji questioned, now sincerely concerned.

Morris threw his hands up. "I told y'all. I told you not to bet on it."

"Guys, look," Jerry tried. "We're on your side. Calvin even tried to deflect some of this and he wound up lying to a federal agent."

"Which turns out," Calvin said, shaking his head, "is a felony."

"Then why his ass ain't in jail then?" Weimer demanded.

"Thanks Morris! I try to cover for everyone and-" Calvin retorted.

"Okay, let's stop!" Jerry yelled, silencing the group. "A lot has been going on while you guys have been in here. Cover up attempts, corporate buy-outs, and federal agents. I know it sounds like a movie but this is for real!

"Now," he continued. "Mr. Maychoff, who is the majority owner of Troyconics, is working closely with the authorities to work out a deal that is mutually beneficial for everyone. But because the government is now involved, things are a little more complicated. We don't know exactly how things are going to shake out."

"Jerry?" Rajin spoke up, his calm voice cutting through the tension. "How long are we going to have to stay here? I have a wife and kids."

"Rajin, I know. You should be able to go home within a couple of days."

There was a groan from a majority of the team, but one of the researchers, Paula Shuf, stepped forward, and bravely asked the question that Jerry had been wondering about since the incident. "When are we going back?"

That shut the team up. Rajin turned to her with horror. "What did you say?"

"I said 'when are we going back?' To the other world. When?"

"I see why you been so quiet," Morris said, dismissing her with a gesture. "You're crazy!"

"You're doggone right she is." Benji muttered.

"Paula, that place is the reason we're all stuck here," Cynthia pointed out. "Why would you want to go back?"

"Because that's the only reason why I came. I loved Lilliana. She was like a mentor to me. But she's gone, along with Johnson and Atkins. Now does not knowing what went wrong still concern me? Of course! But this is what we do. And we have the opportunity of a lifetime here. We can't fold because of the unknown. You can call me crazy if you want, but I want to go back."

"See Jerry?" Morris said with a snort. "She's got quarantine psychosis. She's been cooped up in here too long, you gotta let us out."

The door opened again. Everyone turned to see Maychoff walking in.

"Who the hell is this?" Morris demanded.

"Morris!" Calvin hissed, intending to quiet him before he said anything else.

"You might want to shut up!" Ben said, hitting Morris' arm. "That is Mr. Maychoff. He's only the owner of the whole frikkin' company."

"How was I supposed to know?' Morris said defensively.

"Everyone is free to go… after you give your statements to the National Security Administration agents who should be here shortly." Maychoff exclaimed.

"I knew I liked you. Gimme some," Morris exclaimed as he walked over to Maychoff. He held a closed fist out to Maychoff and to everyone's surprise, Maychoff obliged, bumping fists with Morris, who laughed. "See! My man knows what's up."

Maychoff shrugged. "American TV."

"Where are the agents?" Sara wanted to know.

"They'll be here shortly," Maychoff said in an assuaging tone. "They will meet with each of you individually and ask for your statements. It may take a while, but please be patient. If you cooperate, we can get you home by the end of the day."

"Mr. Maychoff?" Calvin said, nervously shifting his weight. "Shouldn't we make sure everyone has the same story?"

"There's no need, Mr. Madison. Not as long as everyone tells the truth. I already told them everything you and Jerry relayed to me yesterday at breakfast. There's no need to hide any longer."

"Well that's great for all of you," Calvin said, "but I already lied to them. If everyone tells the truth then they'll know I lied and—"

"They knew you were lying as soon as you opened your mouth," Maychoff told him. "I have already corrected that error. It's all taken care of."

Calvin looked over to Jerry, who simply shrugged. "Alright," he said, relieved. "So what do we do now?"

"Give them your statement and go home. I'll work out the details and let you know what's next. Have a good day everybody. And remember: tell the truth."

The team burst into a cacophony of whispers and murmurs as Maychoff left. He didn't try to eavesdrop, even though he knew everyone was talking about him. It was something he had gotten used to. Instead, he pulled out his cell phone, punching in a code to scramble the call as he exited the dormitory.

"*Data set 10,*" a mysterious voice with a thick Ukrainian accent answered. "*Vector 1175.*"

"Data set 10," Maychoff repeated back into the mic. "Counter vector 2286.'

"*Confirmed. How can I help you, Sir?*"

"It appears we have a spy somewhere in our ranks. I need you to flush him out."

"*Certainly. There's a good chance I already know who it is. Give me two days.*"

"Understood," Maychoff said, hanging up the phone as he stepped into the elevator.

Jergensen Residence — Westerville, Ohio, U.S.A.
Earth Date: October 23rd, 2014 — 7:48pm EST

By the time Jerry made it back to his house, he was bone-weary and soul-dry. Between giving his official statement to the N.S.A. (during which he had to describe Lilliana, Johnson and Atkins' deaths in excruciating detail) and coordinating the release of the other team members, he was exhausted.

"Jerry? Is that you?" Candy called from the other room. She entered the kitchen and when she saw him her eyes lit up. "It is! I missed you!"

She crossed over to him and planted a loving kiss on his mouth. "I'll have to leave for a week more often-," his reply interrupted by Candy's extended lip lock. She gives him tongue.

"Hi Daddy!" Alexandria cried out and they broke off their kiss as she ran over and joined the embrace. "I'll pass on the kissing," she laughed. "I think Mom's slobber is all over your mouth."

"Come here little girl," he smiled and pulled her close to him, kissing the top of her head tenderly.

"Daddy! Yuck!" she groaned with faux disgust. She pulled away and wiped the top of her head with her sleeve. "So? How was Mars?"

"What do you mean?" he asked, laughing the question off.

"Hey, you're the one that said you were going to outer space. And besides, you were basically gone long enough to visit another planet."

"You know what, Lexi?" he said, suddenly serious. "You caught me. I actually was on another planet."

"Yeah right," Candy laughed.

"Actually, I'm not kidding. See, I've been working on this device—a Dimensional Mirror—and I couldn't get it to work for a while, but then… Well… I figured it out. I got it to work. And when I finally went through the mirror I discovered a world very similar to our own…"

"Really?" Candy teased.

"Really. It's just not as developed as our planet."

"Dad, you're as bad at lying as you are about leaving voicemails. Anyway, I gotta go."

"Alex! Your Dad just got home," Candy said.

"Mom, I know, but me and Kendra have tickets to that concert tonight, remember? We didn't even think Dad would be back yet."

"It's okay, Sweetie. I'm just glad I caught you before you headed out," Jerry said. "You go have fun."

"Can I have some money?" Alex quickly snuck in, taking advantage of her Dad's mood. Taking out his wallet he handed her a fifty-dollar bill. "Wow, you really have been with the aliens," Lexi said as she folded up the money and stuffed it in her pocket. "They must have done something with your brain."

"You have no idea."

A car horn sounded and Alexandria waved as she headed for the door. "That's Kendra. I'll see you guys later."

"Love you," Jerry said. "Call us after the concert."

He and Candy watched her go, smiling. When she was out of sight, Candy stood on tiptoe and kissed him again, wrapping her arms around him in a tight squeeze. "So," she said as she buried her head against his chest, "how's work?"

Jerry chuckled, stroking her hair with his fingers. "How long do you have?"

"Well our daughter is going to be gone for at least the next three hours... We've got the house to ourselves..."

Sighing, Jerry pushed her away and locked eyes with her. "Candy, you know the story I told Lexi about going to another planet? It's all true! I'm not making it up."

"Jerrrrrry..." she moaned, rolling her eyes smiling. But when she looked back at him and saw his seriousness, she could see that he was telling the truth. She gaped at him. "Jerry... What's going on?"

And suddenly everything was pouring out of him. He told her about his project, about the Dimensional Mirror. He told her about their first mission through the mirror onto the parallel planet. He told her of Lilliana, Johnson and Atkins and how they died. Finally, he explained about the sudden involvement of Maychoff and the N.S.A.

By the time he was finished, they had made their way to the kitchen table, where they sat across from one another in the dimly lit breakfast nook. Candy reached out and took her husband's hand. "I can't believe they would try to blackmail you for doing your job," she said. "They're all corrupt!"

"Candy, I didn't say they would blackmail me," Jerry explained. "But with the cover up and the government possibly buying out Troyconics, anything is possible. I need you to know the whole truth before all of this goes public."

"Well Jerry, do we need a lawyer or something?"

"Honey, no. Calvin and Mr. Maychoff are working out where we go from here, so we'll just have to wait and see."

Candy moved her hand to his face and cupped his jaw in her hand. She turned his head toward her, locking eyes with him. Her face was loving but stern. "Jerry, I'm behind you 100%. This sounds unreal and dangerous. I know you well enough to know that asking you to not do it would be like death to you so I won't do that. You've worked on this for so long and we don't know how it will turn out. But if you ever—*ever*—go to another planet again without telling me… I'm gonna kick you're as-butt."

Suddenly her furrowed brow relaxed and she burst out laughing. Jerry smiled along with her. "Kick my as-butt," he said. "Is that what they're teaching you at church? Spousal abuse?"

"Jerry stop. Of course not. You need to start coming with me."

"Does Lexi go?" he asked.

"No… Not yet. I don't want to force her."

"That's good," he nodded. "I think that's really good. You know I got forced to go to church when I was young. Which I think is the reason I don't go now."

"Don't try to change the subject," she said. "If you leave the planet, you have to at least call me first or—or text me or something."

"Okay Mama."

"I am your mama," she grinned. "Sexy mama!" She leaned forward, kissing him again. Slowly, she opened the buttons of her blouse and peeled it off.

"Whoa!" Jerry grinned.

"What do you expect?" she raised an eyebrow. "You've been gone for a week and a half…"

"Okay, but be gentle with me," he laughed.

She threw him a mischievous smile. "Nope!"

The Pentagon — Virginia, U.S.A.
Earth Date: October 23rd, 2014 — 10:13pm EST

Once again, the cabal of military and government officials were seated around the large conference table. General Maddox sat at the center of the group, facing the largest of the flat screen displays. On the screen, Bernard Maychoff's face was once again projected, however now it was a live teleconference instead of a static photo.

"Mr. Maychoff," General Maddox told the screen, "we appreciate your cooperation and candor regarding this matter. I think we can all agree you've made life easier for all of us. However, we still have a couple of serious issues we'd like to address."

"I'd be happy to help any way I can, General," Maddox said through the live feed.

"First of all, there's still the matter of the attempted cover-up."

"Is that all?"

"Then there's your attempted coup of trying to purchase the remaining ownership of Troyconics without first disclosing your recent discovery to the United States Government."

"General Maddox, I was not aware of the full nature of the discovery when I began executing the buy-out, so for all practical purposes I do not believe it fair to assume criminal intent."

"Mr. Maychoff, I understand exactly what you mean when you say you weren't 'fully aware,' so don't propose to pee on me and tell me it's raining. You knew something. A man doesn't put out almost two hundred million dollars on a whim. There had to have been a reason you were suddenly so interested in buying out the company."

"Your point is well taken, General," Maychoff conceded. "So where do we go from here?"

"Well, quite frankly, I've spoken with my superiors and they fully intend to execute the government's right of first refusal. We're going to buy your company, Mr. Maychoff."

"I see," Maychoff grunted.

"Now, in exchange for the government not bringing any of you up on charges, we are requesting that you allow this transaction to go through without a fight."

"General, I appreciate their offer to overlook any legal transgressions. That being said… There must be a way for us to make this a win-win situation?" He raised his eyebrows.

"You've got some balls on you, Maychoff," General Maddox grumbled with a mixture of irritation and disrespect.

"As do you, General. We both know that even if I allow you to charge my team, and me the world is not yet ready for this type of news. I fail to believe that a headline reading '*Prominent Businessman Arrested for Accidental Deaths on Parallel Planet*' is in the best interest of your country.

"In addition," Maychoff went on after a brief silence, "if I decide to fight this buy-out, my lawyers could stall this deal for a year or more with due process. So again, how do we make this a win-win?"

The General frowned into the camera, pissed. "Give us a moment please."

"Certainly, General," Maychoff said with a smug grin before Maddox slapped the mute button on the conference table's control panel.

The General leaned back in his chair, his face almost a snarl. "That smug son of a—"

"Sir, he's right," said a thin man in a suit, head counsel for the Department of Defense. "If Maychoff decides to fight this thing, he could draw it out for quite some time. He'll have a whole legal team filing injunctions and requesting extra prep time. He has the resources, and he's got the motivation."

Major Williams spoke up as well. "Also Sir, several of Maychoff's business are media companies. We're talking TV news, newspapers, and a couple of popular blogs. He can spin this however he wants."

"The major is right," the N.S.A. chief at the head of the table, muttered quietly. Everyone at the table turned to him, leaning in to listen. "Maychoff is a shrewd businessman. I'm sure he would find a way to leak the story if we threatened him and then we'd have a nightmare on our hands. And if the Russians or anyone else find out about this, things then could get even worse."

"So in other words," the General summarized, "he's got us by the short and curlies."

"Well sir," Williams offered. "He's holding ours and we're holding his. Its mutually assured destruction."

Maddox let out a slight chuckle. "I like that Major. Un-mute him." The major obliged and the General straightened in his seat. "So, Mr. Maychoff. What is your definition of a win-win situation?"

"I have just a few simple requests that I'd like you to consider."

"We're listening."

"First of all," Maychoff said, "I realize that you're entitled to the rights of first refusal. However, the loan agreement also stipulates that you must purchase the company at fair market value. In light of certain recent discoveries, you should know that we now estimate that value to be somewhere around $10.6 billion."

A few of the personnel around the table stirred and whispered to their aides, but the General silenced them with a hand unfazed. "Okay. What else?"

"Well, it should go without saying that me and my team will be given immunity for any perceived criminal acts."

"Yes, yes. Anything else?"

"Yes," Maychoff said. "I am requesting that I remain in place as CEO of the company, and that all of our team members and employees maintain their current positions.

"Wow Maychoff! When you said win-win did you mean for you and us or for you and you?

Maychoff smiled, "Those are my terms. I trust you'll be in touch. Good day, General." And with that Maychoff ended the videoconference. The screens turned black for a second before being replaced with the standby seal.

"He's good," the D.O.D. lawyer said, shaking his head.

"Shut up," General Maddox snapped, shooting a dirty look his way.

Troyconics Astrophysics — Columbus, Ohio, U.S.A.
Earth Date: October 24th, 2014 — 7:56am EST

Well rested and extremely well satisfied, Jerry sat in his laboratory with renewed energy. Though brief, the trip home had bolstered his excitement about the discovery, and had eased his sorrow over the tragedy that had taken place. Now he was newly determined to figure out what had happened and how to solve the problem so that they could go back through the mirror and learn more about the parallel world.

He sat at a worktable that had been erected somewhere amidst the haphazardly constructed server towers. He had his laptop out and was watching the footage of the insertion again. Only this time, he didn't wear the headphones. In fact, he had muted the sound. Instead, he simply used his eyes trying to find some hidden detail that he had missed on his previous viewings.

It wasn't until the seventh time going through the footage, did Jerry see it. A sharp glint of light off in the distance, hidden amongst the trees. Jerry paused the video, zooming in and enhancing the image with a combination of deft keystrokes. There, looming behind a copse of trees at the crest of the valley was a large, castle-like structure.

"You're back bright and early," Calvin said, breaking Jerry's concentration as he sauntered over. "For a man who presumably just woke up, you look beat."

"Well, let's just say that I haven't been home in a week and a half, so Candy was a little horn—"

"Uh, Jerry, that's your business," Calvin waved him away. "I'm just happy to know that your evening escapades mean you won't be trying to kiss me again."

"That was in a moment of joy."

"Anyway. Apparently Maychoff is a pretty good negotiator. We're all of the hook in regard to criminal charges, and we all get to keep our jobs for now."

"That's great," Jerry smiled. "Why the long face then?"

Calving sighed, knowing that Jerry wasn't going to like what he had to say. "The government is taking over the facility. For all intents and purposes we are now under military authority. As of right now."

"What?" Jerry almost yelled. "When did this all happen?"

"General Maddox briefed the President on the situation last night and the President agreed to all of Maychoff's terms with a few caveats."

"So they're executing the buy-out clause."

"Of course they are," Calvin told him. "They'd be foolish not to. However, Maychoff is going to remain CEO. Everything is the same in terms of R & D, but everything we do is going to have to go through General Maddox, and they're going to be stationing some men here at the facility."

"What?"

"It's out of our hands Jerry. This is all a part of the deal. Believe me, it's better than jail and unemployment. But you're going to start seeing a lot of changes around the campus beginning today. Everybody is going to be assigned new security credentials and clearance levels. They'll have to check in and out of the facility at a military guard station."

"Are you serious?" Jerry asked, incredulous.

"Yes, Jerry. Deal with it. Troyconics is now a research and development *military base*."

"What about our research?"

"Everything will continue with a few minor changes, mainly workflow. Jerry, I need you to understand something. Everything—and I do mean *everything*—regarding this project *has* to be reported to me, Maychoff and Maddox now. Every little detail. Every move you want to make. Everything."

"I understand, Calvin—"

"Jerry," Calvin cut him off. "I have never been more serious. No more Lone Ranger. These people play for keeps. Part of this agreement states that if we give any resistance, or are not forthcoming with future discoveries or information, charges can be reinstated and we can still be terminated. This is not just about you anymore. They can replace the entire staff—every single employee—with their own personnel."

Suddenly Jerry found himself with nothing to say. He let the words sink in. "Calvin. I understand."

"I hope you really do. Maychoff has demanded that I make sure you're perfectly clear on this. He's informed me that if I have any doubts as to your ability to comply, I'm to gather all your data and replace you immediately, so I'm asking you as a colleague and more importantly, as a friend. Don't screw this up. There are advantages to having the government involved in terms of money and access to resources. So look at the bright side."

"Yeah," Jerry said, nodding. "Sure."

"We clear?"

"As you say, *Crystal*," Jerry said.

"Good." And then, as if he hadn't even noticed the laptop before, Calvin now looked over Jerry's shoulder at the zoomed-in screen. "What are you working on?"

"What's that look like to you?"

"Uhhh…" Calvin put on his glasses and squinted at the screen. "A castle."

"This is the footage from the mission," Jerry revealed, zooming back out to show Calvin the whole picture. "There's a building there. It's inhabited."

"My goodness," Calvin said, his jaw dropping. "You didn't mention that in your report."

"That's because I didn't see it. I just found it now while going over the footage." He zoomed in again, making the image as clear as possible. "How about that? Looks like there are two people on the balcony."

"People?"

"Or whatever," Jerry shrugged.

"It's hard to tell…"

"Can you get Maddox to approve a return trip?" Jerry asked.

"Jerry—"

"I'm serious. If you can't, maybe Maychoff can. You said he's a good negotiator. We need to go back. I need to find out what happened and see if we can make contact with whoever built that castle—"

"We don't even know if that really is a castle, Jerry,"

"Exactly! Let's go find out! And listen, I think I know what went wrong. See, everyone who died said something."

"What do you mean? Said what? What does that have to do with anything?"

"I don't know," Jerry said. "But I just noticed that certain things started happening after certain things were said. So what are our chances of going back in?"

"You're unbelievable! I'd say slim to none..." Calvin sighed. "But I can ask..."

8:12am EST

Not long after, they found themselves up in Calvin's office, huddled over the black office phone. *"Absolutely not!"* General Maddox yelled at them through the speakerphone.

"General, I understand your concern—"

"I don't think you do, Mr. Madison," the General cut him off. *"Three people have been killed. You and your team attempted to cover it up. The full investigation has barely begun and you want me to authorize an action that could possibly lead to more casualties? I was born at night, Madison, but not last night."*

"General, this is Jerry Jergensen," Jerry said into the phone.

"Well well! The man himself."

"Sir, I believe we have located a structure on the planet and possibly life forms of some kind. There's really no other way to say it. We think they might be able to help us understand what happened. How things work on the planet."

"That wasn't in the report I received," Maddox growled.

"Uh, Sir, we just came into that information this morning, after reviewing the footage from the insertion a second time."

"Madison, let me be clear again. If we find out you're withholding any information, all charges will be reinstated. Your immunity will be null and void."

"Sir, I assure you, we are not withholding anything. The structure was very hard to see, Jerry only found it today, you have my word. It's a castle-like building and there are what look like two figures standing on the top balcony."

"I want that footage forwarded to me on a secure hard drive, ASAP. In the meantime, are you familiar with Major Keira Blake?"

Jerry and Calvin shared a look. "Yeah," Jerry said. "She's a specialist in astrophysics and space-time continuum theories."

"Good! Effective immediately, she is your new research partner."

"Sir, I don't need a partner," Jerry protested.

"I'm not asking Mr. Jergensen. She'll be there first thing in the morning. You will debrief her on all your progress and she is to know everything you do. Is that clear, Mr. Madison?"

"Sir, is all this necessary?" Calvin asked, trying to stand up for his employee.

"Well, the alternative, I'm sure, will not be to your liking. Be sure Mr. Jergensen complies, or I will not hesitate to bring both of you up on charges. Now again, is that clear, or do I need to involve Mr. Maychoff?"

"No sir," Calvin buckled. "We're clear."

"Good. The base commanders and myself will arrive tomorrow afternoon. Please be sure Major Blake is up to speed so that she's ready to go by the time we arrive. Good day gentlemen."

There was a loud *click* as the line went dead. Calvin ended the call. "And it starts…"

"Calvin, this is not good," Jerry said, shaking his head.

"Jerry, you brought it on yourself. If you had just come to me to begin with maybe we could have kept this contained."

"What about Maychoff?" Jerry asked, ignoring Calvin's point.

"What about him?"

"Maychoff has more pull than we do. I don't believe he's signed the transfer paperwork for the buy-out yet. Maybe he can throw in one last condition before it goes through."

"Jerry, you've got to be kidding."

"What do we have to lose?" he smiled. "Maychoff is all about the discovery anyway. That's the only reason he tried to buy the rest of the company. He sees the potential!"

"Jerry…"

"Just let me talk to him."

Calvin thought about it, then let out a long, heavy breath. "Fine. But, if he's not open to it, I want you to drop it. Don't push him or piss him off. Deal?"

"Deal," Jerry proclaimed with a smile. "Is he still in the states?"

"Yeah. He doesn't leave until tomorrow to visit Quilsec."

Jerry furrowed his brow. "Why is he going to London?"

"As if I know," Calvin said with a shrug. "I'll call him and ask for a meeting."

"Great! Thanks, Calvin. You won't regret this."

Rolling his eyes, Calvin picked up the phone again and started dialing the number for Maychoff's assistant. "Sure," he mumbled as he punched in the digits, "where have I heard that before?"

Johnson Air F.B.O. — Columbus, Ohio U.S.A.
Earth Date: October 24th, 2014 — 2:41pm EST

Maychoff's private jet was in the process of being refueled when Jerry arrived at the small airport. Calvin had managed to squeeze Jerry in for a quick meeting at the F.B.O. before the mogul flew off to London. He was five minutes late getting there, which meant he only had another ten to plead his case.

As he walked over to the F.B.O., he saw that there were several military vehicles parked out front, with a couple of Military Police officers guarding the door. When he approached, they stepped in front of him, blocking his path.

"State your business here please?" One of them asked.

"I'm, uh, Jerry Jergensen?" Jerry said in a question. He wasn't expecting to have to deal with anything like this. "I have a meeting here with Mr. Maychoff."

"I.D.?" The M.P. asked, holding out his hand. Jerry handed over his credentials and the officer examined them closely. Then with a curt nod, "They're expecting you. You can go right in."

The M.P.s stepped aside and Jerry pushed through the glass doors of the airport. Inside, more officers were scattered throughout the open space of the airport's atrium. Maychoff stood in the center of the large meeting room, in the middle of a heated discussion with a man who Jerry could only assume was General Maddox.

"We had a deal!" Maddox was yelling as Jerry entered quietly. "Now you're pushing it!"

"General, I believe this course of action is in the best interest of everyone involved. The U.S., the D.O.D., and Troyconics."

"I don't give a damn about Troyconics," Maddox grumbled.

"Which is exactly why I wanted to remain in charge."

"You're not in charge, you're a puppet and I tell you what to do!"

Jerry cleared his throat as loudly as he could, trying to announce his arrival without being too intrusive. The two men turned and saw him.

While Maychoff seemed pleased that he was there, the General had the opposite reaction.

"Ah good. You're here," Maychoff said

"I know you're behind this Jergensen!" Maddox growled at him. "I already told you 'no' over the phone. I don't appreciate these backdoor tactics."

"Tactics?" Jerry said innocently. "What are you talking about?"

"Don't play dumb with me. Maychoff wants to alter our agreement to include a return visit to the parallel planet. This was your idea."

"Actually, General," Maychoff interjected with a smile, "he is unaware of our discussion. But, since he's here now, perhaps he can assist me in explaining the necessity of the action. You see, General, the operation is now under your control. We believe this is essential in pushing the discovery forward. Now that you're involved you can accompany Mr. Jergensen through the mirror and have immediate oversight over the insertion. We simply believe this may help us find out what actually happened and how to prevent it in future visits."

The General thought about it, looking from Maychoff to Jerry and back again as he struggled with his options. Finally he gestured to the M.P.'s, who fell out of rank and began to leave the airport. "I'll have to talk to the President," Maddox said as he followed his men toward the door.

"Good," Maychoff told him."I have full confidence that you will be successful in helping him see the mutual benefit of this excursion."

Maddox stopped in the doorway, looking back at the two men. "This is it, Maychoff! If we get this to fly I want your signature on those papers by morning. No more delays or the whole deal's off."

And then he was gone, the door to the meeting room swinging closed behind him. "What just happened?" Jerry asked.

Maychoff chuckled. "As you may know by now, I am not the most patient man. When Mr. Madison called me and asked if I would meet with you, I insisted he tell me what it was about."

"So Calvin told you that I wanted to go back to the planet?"

"He did indeed. The General and I had other business to attend to, so instead of waiting to meet with you, I brought it up myself. And I think you're right. So, if the President approves, we can go back and visit this structure. And hopefully, perhaps we can make contact with these,.. inhabitants as well."

"That's great," Jerry said.

"If they approve. And even if they do authorize a second trip, there's no telling when they might authorize a third. Perhaps never. So, you must make this trip count. Get the most out of it. What happens will have a great bearing on how things go forward. I must say, you seem very confident about returning to the planet, so I have assume that you have some sort of plan to make sure we do not have a repeat disaster."

"If you're asking whether or not I think there are possible dangers, obviously there are. But I also noticed that nothing took place until certain... until things were said. So far it's just a theory, but I think it's the strongest I've got so far."

"Understood."

"Did they tell you about Major Blake?"

"I'm aware of it," Maychoff nodded. "I had already anticipated such a move. I asked the General to delay her arrival by three days. Hopefully, the trip will be approved by tomorrow and you will be allowed to return to the planet with minimal distractions. However, I am sure the General will insist that a military compliment go with you."

"That's fine," Jerry shrugged. "As long as it's small and non-threatening. We don't want our first interaction with the inhabitants to come off as hostile."

"Yes, but we also need to ensure your safety as well."

Jerry waved away his concern. "I'll be fine. I think I've figured out a few things about the planet. Like I said, hopefully those two people—or 'beings', I should say—can confirm my theories."

"Good. I'll trust your judgment, Mr. Jergensen. See to it that you don't let me down... again."

"Yes sir," Jerry said.

Maychoff checked his watch and then headed for the door. "I have other business to attend to elsewhere as do you. Take the rest of the day off. I'll call you as soon as we're a go. Good day, Mr. Jergensen."

"You as well, sir," Jerry said as Maychoff walked out the door.

Jergensen Residence — Westerville, Ohio, U.S.A.
Earth Date: October 24th, 2014 — 3:37pm EST

As Jerry cruised down the interstate, on his way home, his wife sat at the dining room table reading her Bible. As she read, she took notes in a loose leaf notebook, wanting to be prepared for the Bible study group that met after church every week. She was only halfway through the assigned reading when the phone rang.

"Hello?" she answered.

"*Hey Candy,*" Jerry said. "*How are you?*"

"Fine," she told him. "How's your day?"

"*Uh, great. I've actually just been given the day off. I called to see if you'd like to do something—anything—with your husband today?*"

"Aww, that's sweet honey. I had planned on going to church. There's a guest speaker from out of town tonight. I hear he's really good. Do you want to join me?"

There was a reluctant silence on the other end of the phone. "*I don't know...*" Jerry said.

"Jerry, you said anything."

"*Well yeah, but I wasn't thinking church. I was thinking more sex, maybe dinner...*"

"Well that's perfect. You can come home now, we can have sex, you can take me out for an early dinner, and then come with me to church at 6:30. There's plenty of time. That way you get what you want and I get what I want. Deal?"

"*Uhhh...*" was all Jerry had time to say.

"Great! It's a date! See you soon, Honey," she said with a smile before hanging up on her husband.

4:37pm EST

Only an hour later, Jerry and Candy were locked away in their bedroom, from which the moans and grunts of pleasure were audible, even through the closed door. "Oooohhhh," Candy could be heard from the hallway. "Oh boy! Oh boy! Oh BOOOYYY!!!"

Inside, they shared a climax that sent both of them into involuntary spasms and bathed each of them in the brief euphoria of carnal satisfaction. Finally, Jerry rolled off of his wife and the two of them lay beside each other, naked and panting.

"Oh my goodness," Candy breathed. "I haven't come that hard in years!"

Jerry couldn't help but smile as he caught his own breath. "I do my best," he told her.

"There are so many women out there who would pay a lot of money to have the orgasm I just had," She said, absent mindedly reaching out and stroking his thigh with her fingertips.

"Good. It was good for me too, but now I'm really hungry…" He rolled over onto his side, draping an arm over Candy's slim waist. "You know, I was thinking. Maybe we should just order in some Chinese or something. Just eat here, you know?"

"That's fine with me," she said, still hypnotized by the aftermath of their antics. "As long as you go get it."

"No problem," he told her with a kiss.

There was a few seconds of silence, and then she rolled on her side as well, looking him in the eyes, suddenly intense. "Jerry? Why did you agree so quickly?"

"What do you mean?"

"I mean you hate going back out once you're home. Unless it's for work. What are you up to?"

"I'm not up to anything," Jerry stammered defensively, invisible hives of guilt breaking out across his entire face.

"You really don't want to go to church, do you? You know I get sleepy after I eat Chinese food, and if we're already at home I'm not going to want to go. Jerry, that's so underhanded."

"Honey, the church thing is all you. I believe, but—"

"Jerry, that's not the point. You already agreed you'd go with me. Now we're going out to eat."

"What?"

"It's your fault. Come on. Get cleaned up." Then she rolled off the bed, out of his light embrace. She headed for the bathroom and the shower within.

"That was supposed to work another way," Jerry grumbled as he lay on the bed.

"What did you say?" Candy asked before closing the bathroom door.

"Nothing, Honey!" Jerry told her, slapping his chin in self-punishment. For a brief moment he felt like he was a child again with his father who made him attend church but he had already given Candy his word.

House of Grace Christian Center — Westerville, Ohio, U.S.A.
Earth Date: October 24th, 2014 — 7:15pm EST

By the time they arrived at Candy's church, the service was already well under way. Entering through the double doors in the back of the filled seats, Jerry followed Candy down the aisle to what he presumed was her regular seat in the fifth row. The congregation, which was currently on their feet, were just ending a beautiful song, failed to notice their arrival— or if they did, they at least didn't make a show of it—and soon he found himself being handed a service program by his wife.

"We're so late," she whispered under the loud singing as she opened his program to the order of service. "We missed the praise and worship…"

"Who knew it would take so long to get our meal," he muttered to her. "I told you we should have done take-out."

"You had ulterior motives."

"Okay," he shrugged. They were here now, and it was no longer worth arguing about.

"At least we didn't miss the speaker. He's supposed to be really good."

"Let's hope he can keep me awake," Jerry said under his breath.

The Pastor was now up and quickly introduced the evening's guest speaker. Jerry's mind was already wandering, thinking about the Dimensional Mirror, the parallel planet and all the unanswered questions he had, when a handsome and distinguished man in a well-tailored suit approached the podium.

"What if I told you and showed you by the Word of God," the minister called out to the congregation. "that all life on this planet is word originated and word controlled? All of it!"

And suddenly, he had Jerry's attention.

The minister went on. "When God said 'light be'— when He released those words, He also released faith in those words to bring what He said to pass. Think about this. The Bible says that God is *LIGHT*! Think about that for a moment. God is Light! So when He said, 'light be' he was basically saying what's in Me BE! So for all practical purposes God was simply releasing Himself. The universe declares His Glory.

Imagine a painter or artist, with a canvas that just keeps getting bigger and bigger and bigger. The more he paints the bigger the canvas gets.

That's how the universe is and on that day, it began expanding at the speed of light and it's still doing it today. Scientists can confirm that the universe is still expanding. Why? Because God said 'light be, light went', and He never told it to stop. And then he furnished it… he filled that canvas with all of His creations. But it all began with two words: *'light be.'*

"So what does this have to do with life and words: 'in the beginning was *the* Word, and the Word was with God and the Word *was* God. All things were created by Him, by the Word. And so everything else from that moment of 'light be' on was geared to be created and controlled by words. Now after He created the Earth with His Word, He then with His Word gave control of the Earth to be under the dominion of human beings. He said 'let them have dominion.'

"What you need to see is that light or light energy is in everything. Scientist know this. Light is the component, the foundational component that God used to create everything He created with words. That is how He operates in creation. And everything, all things, all stars, all planets, all material, The Word, His Word created all that.

"The Word was the mechanism God used to take it from the unseen realm to the seen realm where you and I live. Now if you can put yourself at the beginning with God and He didn't tell you what He was doing you simply saw it, you would probably ask, 'why is He doing this or what is He doing this for?" And both of your questions would be wrong. Not what but who? Because just like a mother preparing a room for her baby God put everything on this canvas piece by piece, step by step until all was ready then He said 'Man, be in Our image." He created this entire universe for us, for man. But again every piece of creation even man himself was created by *words.*"

At this point, Jerry was leaning so far forward on the edge his seat that, he was close to falling off. In fact, he very well may have, had Candy not motioned for him to move back. "Are you okay?" she whispered.

Jerry didn't answer. He was staring at the minister with fascination and intrigue. It wasn't until she grabbed his sleeve and shook his arm that he responded. "Huh?" he said in a hushed voice. "Yeah, just some interesting concepts."

"Concepts? Jerry he's teaching from the Bible. It's not a concept."

"What do you mean? He hasn't brought up one scripture."

"Oh Jerry, he's talking straight out of the first book of Genesis. Everybody knows that—"

"Shhh," he quieted her, turning back to the message. "Whatever, it's interesting."

"How does man create things today?," the minister continued. "He comes up with an idea or a picture of what he wants or thinks. Then he shares that idea through *words* and turns it into a blueprint or design for what he wants to create. Where did man get that ability from? He got it from God! God created everything through words and pictures in His own mind. He then speaks the words or sounds with that picture in or on His mind and He has to power to bring His own word to pass or manifestation. Man does things the same way… it just seems to happen and work a lot slower. Everything you see in creation is what God had already planned and established inside Himself *including you*! He knew what a zebra would look like before the zebra was ever manifested on the earth. He knew what a peacock looked like before the peacock ever manifested right down to every color and tint of every feather. He knew what all of *this* looked like before He ever said '*light be.*' It was all a very deliberate plan brought into being by His Word!

Jerry reached into his pocket and took out his cell phone. He launched a notepad app and began typing furiously with his thumbs. He wanted to jot down everything the minister had to say.

However, almost as soon as he began taking his notes, the phone began to vibrate in his hands, ringing loudly with obnoxious electronic chimes. Candy scowled at him and heads turned toward them as he fumbled to switch the phone to silent. Embarrassed he glanced at the screen: *Bernard Maychoff.*

"Sorry," Jerry whispered to Candy, who was staring at him in disbelief. "I gotta take this." He stood and slid his way out of the aisle crouching as he walked back down with one finger up and headed out the doors. A moment later he was walking down the front steps of the church as he answered the call. "Mr. Maychoff."

"*Mr. Jergensen. Everything is a go. You and Mr. Madison will meet General Maddox and the rest of your insertion team tomorrow afternoon at the facility.*"

"What do you mean 'insertion team?'" Jerry asked, pacing back and forth on the sidewalk.

"*Unfortunately, that was one of the compromises we had to make in order to get him to agree to allow you to go back in. An Army team has to go with you.*"

"Mr. Maychoff, with all due respect—"

"*Careful Mr. Jergensen,*" Maychoff warned, "*I have not yet recovered from the 'gentle treatment' I received from your government while securing you access back to the planet.*"

Jerry persisted. "Sir, I appreciate that—"

"*Then shut your mouth and take it up with the General in the morning.*"

And then the line went dead. Jerry let out a heavy sigh, frustrated with how his affairs continued to be meddled with by men that did not understand the nature of what they were dealing with. Suddenly, he remembered the minister and ran back inside to catch the end of the message.

Chapter 5

As Jerry entered the lab that morning, it was now infested with almost two dozen military personnel, all of them prepping for a stroll on the parallel planet. General Maddox was walking through the ranks, checking on his men as they geared up. Jerry followed close behind, trying to keep up amid the hustle and bustle of the grunts.

"General Maddox!" he called after the gruff military man. "The main reason I asked to go in alone is to prevent any further loss of life."

"Geez, give 'em an inch and they take a mile…" Maddox grumbled. "Mr. Jergensen, I've had about all I'm going to take from you and Maychoff. No matter what we do, you both keep asking for more."

"There are unique circumstances that increase the risk of possible death. The more people that are added to the team, the chances and likelihood that someone else will die are only going to go up."

"What are you talking about?"

"You wouldn't believe me if I told you, General."

Wheeling around, General Maddox bent so that he was eye to eye with Jerry. "Try me."

The soldiers were all watching now, waiting to see their General tear into the pestering scientist. Jerry said nothing. "Well? Asked the General."We're all waiting."

"I… I don't know how to explain it to you without sounding crazy, but—"

"Madison!" the General cut him off with a bark, calling to Calvin who was somewhere in the back of the lab, talking with a handful of engineers. "What kind of Mickey Mouse outfit are you running here? First the guy tells us we are in danger of dying then says he doesn't know how. If you don't have a psych team evaluator, we've got plenty to spare. In fact, I strongly recommend a full psych evaluation immediately."

"If your team goes, I won't," Jerry seethed through gritted teeth.

"Jergensen, you're in no position to play chicken here, but need I

121

remind you that we can reinstate all charges and have your team removed immediately."

"Make my day!" Jerry yelled, finally daring to raise his voice. "Let's see how far you get with this project without me."

"You're pushing it, Jergensen!"

"No you are!" Jerry shouted back, refusing to stand down. "Your men could die and I already have enough blood on my hands. So what's it gonna be?"

The General stared at him his cheeks flushed red with rage. He snarled. "Alright Jergensen. No team. But you're going to take a single man with you."

"I said alone!" Jerry insisted.

"I give the orders around here now, not you!"

"I said alone!"

"Jerry!" Calvin spoke up, finally walking over and getting between the two men. "It's one man. Let it go."

"Calvin," Jerry pleaded. "You know—"

"I'm not asking," Calvin said. "I'm telling you." He turned around to face Maddox. "General, we'll accept the one man escort and thank you for your kind consideration."

The General growled and stalked off to re-brief his team. Calvin held out his arms to Jerry as if to say *what are you doing?* but Jerry turned away. "I thought we were on the same team!"

"We are," Calvin told him. "But what you don't understand is that your lack of being forthcoming with information has caused them to not trust you, especially where they can't see you and I can't say that I blame them."

"Calvin—"

"Jerry, the General is right. You've been barking around down here giving orders to everybody and making demands. Now it's time you take some. Get yourself prepped for insertion by three o'clock and not another word between you and the General or I'll fire you myself. There are more people's lives and freedom involved here than just yours. You can't possibly speak for everybody."

"I can and I will if it means making sure more people don't die."

"Well Jerry," Calvin snapped, finally fed up. "You can do what you want, but if you don't show up for prep, you're going to see my more adamant side. I hope I'm making myself clear."

"Crystal!" Jerry said still in defiance.

"Good," Calvin said, and walked away, leaving Jerry there to steam alone.

2:57pm EST

After careful consideration, Jerry had come to the conclusion that they had finally backed him into a corner. The government owned Troyconics now. The Dimensional Mirror belonged to them and they were going to do with it pretty much whatever they wanted. The best he could do was take the one man and be happy he had been able to get a trip back to the planet.

And so, three minutes early for the prep meeting, he entered the conference room to find Calvin already sitting at the table with General Maddox and a young man in a Sergeant's uniform.

"Jerry," Calvin said without a trace of a smile. "So glad you could join us."

Jerry matched his frown as he nodded. "Good," Maddox said as though they hadn't been screaming at each other earlier that day. "You're just in time for prep. Jerry Jergensen, meet Sergeant Marc Barclay."

Sergeant Barclay stood and Jerry held out his hand to shake, but Barclay saluted him instead. Jerry returned the gesture awkwardly and then sat down at the table with them. "Good to meet you," he told the Sergeant.

"You as well, sir."

"I'm not a Sir," he said with a frown. "Jerry is fine."

"Sergeant Barclay is a weapons specialist," Maddox explained, "as well as an astrophysics major. He's been out on at least twelve inner-planetary missions and at least thirty-six virtual simulations. So he'll be able to assist you on multiple fronts. Against my better judgment, the President has placed this mission under your authority Jergensen. In the event that there is trouble, that authority automatically switches to Sergeant Barclay. Now is that clear Mr. Jergensen?"

"Crystal," Jerry said, shooting Calvin a look.

"Good," Maddox said. "Now, I've partnered each member of your team with a member of our team in the lab."

"Planning a takeover, General?" Jerry asked.

"I don't know why I'd have to take over something that already belongs to me, Jergensen. No, it's not a takeover. I'm just big on redundancy and I like being prepared. It's simply a wise move for all of us as a team to have multiple people who can perform the same function. Just in case something should happen to one of them, we have another to take their place. Now, is there anything else that we need to know before you head out?"

"I don't think so," Jerry answered. "But talk to Calvin, he's the boss."

"Wow, Jergensen. That's the first time you've deferred command since I met you. You may turn out to be alright after all." The General's words were clearly sarcastic, but Calvin shot Jerry a smile all the same. "Well gentlemen," the General continued, "if that's all, I believe its time for you two to get suited up."

"Sir, on the last insertion the air was very breathable—"

Maddox turned back to Jerry and silenced him with a look.

"But better safe than sorry, I suppose," Jerry finished, understanding that this was a battle he would not win. Instead, he led the others out of the conference room and took them back down to his lab to get ready for another trip through the mirror.

Quilsec Astrophysics — London, U.K.
Earth Date: October 25th, 2014 — 9:46pm GMT

On most nights, Charlie never saw any of the Quilsec employees. He was a custodian, part of a cleaning service that came to the offices after hours to vacuum, tidy and take out the trash. Usually everybody had already gone home and Charlie could work in peace. Tonight however, someone was still working. Outside the door to the big office at the end of the hall, a large man stood guard with statuesque stoicism.

"You can skip cleaning this one tonight," the man said smiling when Charlie approached, his cleaning cart squeaking as he pushed it forward.

"Okay," he said. The large man didn't look like someone worth arguing with. And besides, Charlie thought, he had just been saved a considerable amount of work. Perhaps now he could get out early.

Inside the office, two men who were completely unaware of Charlie's existence sat on either side of a colossal desk. One of the men—the one in the visitor's seat in front of the desk—was Bernard Maychoff. The other man, who sat behind the desk staring at a laptop screen, was tall, sinewy, sported a distinct goatee and wore expensive designer glasses. He spoke with a thick Ukrainian accent.

"How was your flight?" he asked Maychoff.

"Long," Maychoff replied. "Thanks for adjusting our meeting time. I had a few other things come up that required my attention here."

The Ukrainian man continued to talk, simultaneously typing away on the laptop at full speed. "Not a problem, Sir." Finally, he slid the laptop around to show Maychoff the screen. "This is footage from one of our offices here at Quilsec. As you know, each office smoke detector is also equipped with a small camera and audio device."

"Yes," Maychoff retorted. "I don't like surprises unless they're coming from me."

The Ukrainian man pressed the space bar and the footage began playing. On the screen, Maychoff saw a young employee enter an empty office space and answer a cell phone.

125

"The man in the footage is Brian Ridley," the Ukrainian narrated for Maychoff. "He's been with Quilsec now for four years, and works in the Research & Development department. After checking him out, his resume reads like a movie character. Everything is just too perfect. He graduated at the top of his class from M.I.T., a few stints at NASA, even a gig in Area 51 of all places. Then six years in the military with action in Kuwait and Iraq. I'll let you listen to the rest."

He stopped talking and let the footage play out. In the video, Brian Ridley had just begun to answer a phone call. "*Hello?*" he said into the phone. There was a silent pause as he listened, then: "*Yes. Who is this?*"

Maychoff watched closely as Ridley listened intently. When whomever he was speaking with finally finished speaking, Ridley said, "*Go on,*" and grabbed a pen off his desk to take notes with.

"I don't understand," Maychoff said. "We can't even understand what the other person is saying—"

The Ukrainian held up a finger, indicating to Maychoff there was more to come. That's when Ridley said it: "*That's not up to me, Mr. Hankins. That's up to my superiors. They'll have to evaluate how valuable the information is.*"

"Hankins…" Maychoff muttered in surprise as the Ukrainian slapped the space bar again, pausing the footage.

"I believe so," he said, his accent full and thick as ever. "As you can see, this call took place shortly after you spoke with Mr. Hankins. So, it appears that Hankins is playing both sides of the fence."

"What do you mean?" asked Maychoff.

"I did a bit of digging. Apparently Mr. Ridley was recruited by the C.I.A. shortly after his military tour in Kuwait, and was placed as a covert operator in the Army for a few years. That was his specialty. He's normally used as an insert into places the C.I.A. needs to keep tabs on. He's such a brain that nobody pays him any attention, except us. He's great at his job. All of his performance reviews are flawless."

Maychoff nodded slowly, processing all of this new information. "I'm grateful that he is an exceptional employee, albeit a deceitful one."

"What would you like us to do?" the Ukrainian man asked.

"Let's put a bug on the bug."

The Ukrainian let out a hearty laugh. "My pleasure. The usual places?"

"Yes," Maychoff said. "And then some. Add his car—video and audio, of course—and have his cell phone cloned. I desire to know what is the C.I.A's interest in me. After all, I'm not a criminal."

"Maybe they know something you don't," the Ukrainian suggested as he slid his laptop back around so that it faced him.

"Perhaps. Good job, my friend. Can you get in touch with Patricia?"

"The senator?"

"Yes," Maychoff nodded. "She sits on the intelligence review board and has many contacts in the C.I.A. She may be able to help me get to the bottom of this."

"She's also running for President," the Ukrainian pointed out. "Her every step is more than likely under a magnifying glass right now. Do you think it's wise for us to contact her at this time?"

"I should hope so," Maychoff grinned, "seeing as how I'm a major contributor to her campaign."

"As you wish."

"Thank you for your caution," Maychoff said, standing up from his chair. He reached into his pocket and produced a small black velvet pouch. He handed it to the Ukrainian man, who loosened the cinch and spilled the bag's contents out into his open palm. Against the rough skin of the man's seasoned hand, twenty uncut diamonds glittered under the fluorescent lights. "You never did tell me why you prefer diamonds to cash," Maychoff mused.

"Ah, my friend," the Ukrainian man said. "You have your secrets. I have mine. But I will tell you this: they're much lighter to travel with."

Maychoff smiled. "I'll be in touch."

"As will I."

Troyconics Astrophysics — Columbus, Ohio, U.S.A.
Earth Date: October 25th, 2014 — 4:03pm EST

Jerry stood in front of the Dimensional Mirror with Sergeant Barclay at his side. Both men were clad in their environmental suits, and Jerry could hear his own breath echoing in his helmet. It had given him chills to put the suit back on, as the memories of the last mission continued to flare up in his mind like painful goose bumps breaking out on skin.

"Now," Mario, who was reviewing the atmospheric scanner with Barclay, instructed. "Make sure that you double check the nitrogen content on top of making sure the oxygen mixture is safe."

"Thanks," Barclay said. "I've got it."

"That's right," Mario laughed. "I keep forgetting you've done this before."

"Only a few times," Barclay replied, friendly enough.

"Alright!" Maddox barked from the central control console, where he stood with Ian, Rajin and Calvin. "Insertion in two minutes!"

As the rest of the scientists who would be monitoring them from the lab milled about, making their final preparations, Calvin caught Jerry's eye and walked over to him. "Thanks for making the adjustment," he said.

Jerry shrugged. "I guess I should just let you be the boss... Sometimes."

"Don't start," Calvin chuckled. "Be safe."

"Insertion in one minute, thirty seconds!" Maddox updated the room.

"Oh crap!" Jerry suddenly realized. "I almost forgot! Calvin, can I borrow your cell phone?"

"What, now?"

"Please," Jerry said, locking eyes with his friend and relaying with nothing but his expression how serious he was.

Finally Calvin acquiesced and handed Jerry his phone. It was hard to punch in Candy's number through the fat finger-pads of his suit gloves, but eventually he made it. He pressed 'CALL' and put the phone to his ear.

"Hello?"

"Hey Baby!" Jerry said.

"*Jerry, is everything alright?*"

"Absolutely. Look, I can't talk for very long. I just wanted to let you know that I have to… I'm leaving the planet again."

"*You remembered,*" she said with a smile in her voice.

"Of course I remembered," Jerry said, deciding not to tell her how close he had come to forgetting. "I love you."

"*I love you too,*" she told him. "*Be safe.*"

"I will. Bye Honey."

He had barely hung up and handed the phone back to Calvin when he heard Maddox yell again. "Jergensen! On the platform!"

"Be safe," Calvin told him, giving his shoulder a supportive squeeze.

Jerry saw that Barclay was already waiting for him at the bottom of the mesh metal gangplank and he went over to stand next to his military escort. "Ready?" he asked the Sergeant.

"Born ready," Barclay said.

Jerry nodded and soon the *whir* of the motors became a roar and the Mirror Blades were spinning in front of them, faster and faster until—

"Power's up!" Ian called out behind them. "Activating laser light!"

In front of them, the red laser light flooded the center of the spinning mirror mobile and bubbled out from the center as it always did. Then it flattened and opened onto the lush valley of the parallel planet.

"We have a stable portal!" Rajin yelled over the noise.

"Well," Jerry said with a shrug. "There are no flames popping out of it. I guess that's a good sign."

"Gentlemen!" Maddox yelled to them. "You are a go. Godspeed."

And with that, they began walking up the gangplank, the metal clanging with echoes of finality as they made their way closer and closer and then *through* the Mirror. Then they were gone.

Dimensional Mirror Insertion Point — Dabar
Date & Time Unknown

What Jerry found on the other side of the portal, was the last thing he had expected. The last time he had seen this place, it had been consumed by fire and was a holocaust of heat, combustion and destruction. Now, the valley looked as if it had never been touched. The ground was green and covered with healthy grass. There were no scorched trees or charred patches of dead land. Everything was just as it was when they had first arrived.

"How can this be…?" he wondered out loud as he and Barclay stepped away from the portal.

Barclay looked at the atmospheric scanner in his hands and studied the data read-out. "Atmosphere is reading as safe. I thought you said there was a forest fi—"

"Sergeant!" Jerry shouted, whirling on him, eyes wide as he removed him helmet. "I need you to be quiet and listen to me. This is very important. There are things you do not know about this place. Things I could not tell General Maddox because he would refuse to believe them."

Seeing no harm came to Jerry from breathing the air, Barclay stared at the man who had led him to this new world, torn and reluctantly, he removed his helmet as well. He was a military man, and he was loyal to the chain of command. But this was clearly bigger than anything the General—*any* General—had ever dealt with. He decided to wait till he knew more about their situation before choosing a side. He took a deep breath in. "This place looks like Earth."

"Sergeant! *I need you to be quiet!*" Jerry grabbed Barclay shoulders and forced him to make eye contact. "I need you to do everything I tell you while we are here. You have a wife don't you?"

"Yes," Barclay said.

"And you have a son?"

"Yes."

"Do you want to see them again?" Jerry asked.

"What do you mean?"

"Do you want to see them again?"

Barclay looked around, uncomfortable. "Yes, of course" he said.

"Okay," Jerry nodded, loosening his grip on the Sergeant's shoulders. "This is going to seem like a weird order, but I need you to follow it to the letter. As if General Maddox gave it himself."

"Okay—"

"*Shhh!* Do not speak. Do you understand?

Barclay nodded.

"Good," Jerry continued. "Trust me. This is for your own safety." He let go of Barclay and turned around, pulling out an intense pair of binoculars while scanning the tree line. "The structure from the footage is there," he said, pointing to the ridge of the valley ahead. "Just behind the tree line. We estimate it to be about two and a half miles away. It shouldn't take us too long to get there. Come on."

He motioned with his arm for Barclay to follow him and the Sergeant silently obliged. Together they began their trek across the eerily familiar alien landscape, slowly trudging through the grass and into the rejuvenated forest.

Climbing up the hill of the valley was difficult in the suit, and Barclay had to remind Jerry to pace himself. Finally however, they made it to the top, and got their first view of the horizon. The parallel planet stretched on forever, vibrant and alive with overgrowth. Off in the distance Jerry could see a sparkling blue river and could even make out the gray granite palisades of a waterfall.

And on the other side of the rise was the structure. Just as it appeared in the footage, the building looked just like a medieval castle, it's stonewalls and towers standing as a monument to engineering. It was the only man-made (or perhaps *alien*-made) structure in sight, but it sure was an impressive one.

As they approached the front of the castle, a darkened archway came clearly into view. Sensing movement behind the archway Jerry and Barclay to stopped in their tracks. The archway appeared to be pitch black with no door, but suddenly light began spilling out from behind and a silhouette figure began to emerge and continued forward until the being was standing before them.

As the figure stepped forward, Barclay reached for his sidearm, but Jerry put out a hand to stop him. "No, no. Put it away. It's alright."

"But Sir—" Barclay started.

"*Ssshhh!*" Jerry silenced him again and pushed away the man's hand from his holster.

"Greetings, Jerry," an ethereal voice emanated from the silhouette as it came closer to them. When it was only ten feet away, it stepped into the daylight, its features suddenly becoming clear. It was humanoid. In fact, by all accounts it appeared to be a seventeen year-old human male, but there was something different… something glowingly tranquil and more ancient than seventeen. As though aware of everything and nothing all at once.

"We have been expecting you," he said in an echoing tone.

"How do you know my name?

"You and your team were none too quiet on your last visit," said the being, ignoring his question. "And of late we receive so few visitors. You may follow Us. Your companionhowever, must stay here."

Jerry looked back at Sergeant Barclay, who opened his mouth to protest. "Sir, I don't—"

Shaking his head adamantly. "I'll be fine, Sergeant. Please wait here.

Barclay complied with the order, and seconds later, Jerry found himself following the familiar looking life form toward the massive castle and the light that poured out of the previously darkened archway. He took a deep breath as they neared the entrance and held it as they entered what appeared to be a room. The light in the archway once again shifted to complete darkness. Sergeant Barclay stood unsure and alone.

The Sanctuary of Us — Dabar
Date & Time Unknown

The inside of the castle was a labyrinth of broken physics. The clean marble floors were walls and the marble walls were floors and gorgeous antiques furnished every twisted angle of the bizarre architecture. Staircases chased one another as though in an M.C. Escher painting and yet no matter where Jerry followed the being, his feet stayed planted firmly on the ground and he experienced no nausea or disorientation.

The being took Jerry through the room and out onto a terrace that looked out on the parallel planets vast, untouched landscape. Here on the terrace, with the horizon acting as a guide, Jerry knew that the terrace was right side up.

The being motioned to a beautifully carved table and two chairs that were waiting besides the railing of the balcony. Jerry sat down and the being sat across from him. Jerry had so many questions for it…him. "Is there a reason you wanted me alone?"

"It is not about what we want or what you want," he told him in its strange voice. "But rather, it is about what is necessary and appropriate. The Sergeant's heart is not true, and the things we need to share with you would be more than likely be abused by him and his superiors. Abuse always leads to death… Here, very quickly. On your world? Very slowly."

"You know of our world?" Jerry asked, amazed.

"I was from your world," the being said. "I was once a man like you, and in some ways… I still am. But for the now, *We are Us.*"

Jerry cocked an eyebrow, confused. "We are us? Who is we?"

As soon as he was finished speaking, more of the beings poured out of the castle onto the terrace, seemingly gliding across the balcony floor and enclosing their table in a semi-circle against the railing. There were eleven, all but one of them looking just as young as Lexi, all staring intently at Jerry in on their own sense of calm.

"*We are Us!*" the beings all said in unison, their dozen voices echoing together in a beautiful chorus.

"I am the ThreeofUs," the being seated across from Jerry said. Allow Us to introduce you to the other eleven of Us!"

Jerry looked at their strangely perfect looking human faces, speechless from wonderment and curiosity.

"We do not expect our titles or names to make sense to you," ThreeofUs continued. "As your world is rooted in selfishness and individualism. We have since shredded our pride and know who we are as individuals, and we realize even more so who we are as *Us*.

"As it appears that it is yours and your people's intent to continue visiting this place, AllofUs have agreed to provide you with some basic information as how to function here. We will not tell you everything as that would be counterproductive to your growth, but we will attempt to circumvent any additional loss of life such as experienced by your earlier colleagues."

"You saw that?" Jerry said, slightly perturbed. "You saw that? You saw what was happening and did nothing? Why didn't you help when you saw my other teammates dying?"

ThreeofUs closed his eyes for a moment, and as if almost on cue, the other eleven had now disappeared back into the terrace entrance and through another pitch black corridor at the end of the room. Jerry looked around for any trace of them, but found none.

"Jerry," ThreeofUs said, "Unless something has changed over the last few years, you are aware as we are that human beings are by nature hardheaded, stubborn and unwilling to listen to instruction, *believing only what they can see!*"

The ThreeofUs continued on in his passive echo. "Had we told them what to do to cease the carnage they would not have listened to Us just as they did not heed your incessant commands to still their tongues."

"Where were you?" Jerry demanded. "How did you see all this?"

"Jerry Jergensen, although the loss of your teammates is sad, until you are able to see past your offense at our inaction, we cannot share more with you."

Jerry thought about it, breathing deep. Finally he nodded. "Okay. No offense taken."

"Good," ThreeofUs said. Gesturing with a celestial robe covered arm, showing off the endless landscape. "We call our world Dabar. It is a world much like Earth, although as you can see, it is not as well developed or built out. Neither was Earth… in the beginning. But that time for you is long passed.

"Dabar, however, remains rich in resources, and the building of things here is a simpler task than it is on your world. Depending on one's viewpoint. Everything on this planet is controlled—directly or indirectly—by words. Everything on this planet *is* words! Do not forget We have shared this with you, Jerry Jergensen. It will be your first and most important lesson.

Suddenly ThreeofUs reached out and touched his long, slender index finger to the breast of Jerry's suit. Only he didn't just touch it, his finger went through it, almost to the last joint. In fact, his finger went through Jerry. He could feel it slip through his chest and into his heart. Jerry looked down, shocked and dazzled, then by instinct he stepped back slightly.

Then ThreeofUs removed his finger, and continued speaking as if nothing had happened. "On your world, particularly in your country, freedom of speech is deemed to be extremely important. The ability to say what one thinks and wants, to whomever they want, wherever they want, whenever they want. No matter if it is positive, negative or offensive.

"In most cases on your world, this is done with impunity and no apparent repercussions. Here on Dabar, freedom of speech is true freedom. The freedom to create or to do destroy. Here, freedom of speech is the freedom to die or the freedom to live, the freedom to be sick and weak, or to be strong and renewed on a daily basis. At what you refer to as 'mature ages,' people on Earth believe they then know better than all, and believe they can say anything, as we mentioned, without apparent cost.

"Here, every word costs, and is rewarded with either life or death, good or bad, blessing or one could say cursing. And reminds Us to deal separately with what so many of your people have become accustomed to and comfortable with… tell Us, do they still refer to it as profanity or has it simply become a part of the English language now?"

It, uh…" Jerry said. "It's still called profanity by a few people…"

"Ahhh," ThreeofUs smiled. "So you do not use it. Good. This will make your transformation much easier."

"I'll be honest," Jerry admitted, "I've slipped a couple of times. I mean who hasn't?"

"More than you know," ThreeofUs said with a stare that looked right through Jerry. And with that he stood and entered the sanctuary once again, continuing his explanation. Jerry followed him back in, intrigued. "Here on Dabar, many words are profane that are considered normal speech on earth. Specifically words associated with death, hell, fear, negativity, sickness, disease, and many other undesirable texts.

The key is, it is not that some *Being* is dictating morality, but rather it is simply definition and actuality.

"Words here are designed and used and taken and manifested in their actual literary sense or definition, meaning 'bad' is indeed bad and 'good' is indeed good. 'Hot' is considered hot and 'hot as hell,' as one of your colleagues said… is indeed that hot."

Jerry eyes silently expressed his comprehension as he slowly began to digest what was being shared with him. "So, you've just spoken these words. If what your saying is true, how come nothing is happening to you right now?"

"As the ThreeofUs, we are a teacher," the being explained. "However, it is not our title that shields Us from disaster, but rather this sanctuary. Certain areas have been deemed by *Substance* as… you might call them classrooms. They are what we call *Grace Asylums*. However, nothing can be created here with words, good or bad, because of the Grace. In an effort to extend that grace in a limited measure, we were allowed to set up the small metal devices that are strategically located across Dabar."

"The six second boxes…" Jerry muttered, understanding now crashing down on him like a tidal wave.

"That is an acceptable name for them," ThreeofUs said with a slight nod. "After a person speaks a detrimental, damning or negative word toward themselves or others, they then have six seconds to retract or counter those words or they will receive the corresponding actions of those words."

"Who makes these decisions?" Jerry wanted to know. "Who governs all this?"

"Each person makes their own decisions," ThreeofUs said simply. "The law is already in place. Each person chooses what he or she receives by their own words. The six second boxes, as you call them, are a form of temporary Grace."

"Not if you don't know what they are!"

"You figured it out, did you not?"

"Yes," Jerry nodded. "At the cost of three lives."

"Believe Us Jerry Jergensen, that knowledge once cost Us as well. Have you been able to capture all We have said on your recording device?"

Jerry's eyes went wide. How did they know?

"Do not be surprised, Jerry," ThreeofUs reassured him. "We were aware of the device when you activated it before we entered the sanctuary... A somewhat rude but permissible action as We can see you meant no malice. We have given you enough to get started. A man of your understanding will be able to fill in the gaps and assist your colleagues, countrymen, and others on your world in keeping themselves alive while here, if they will take heed to your words. Goodbye Jerry. We are sure you will see Us again. Soon!"

"As he turned and proceeded to step away he paused, "Did you have a question?"

Surprised but without hesitation he asked, "You said you are a man but you look like a young boy....", Jerry probed awaiting an explanation.

"Yes, a surprise benefit from extended time spent in Dabar's atmosphere. The more one spends time here; their youth is renewed, if allowed ... even to their teenage years eventually. Even a short period of time here can prove, physically beneficial."

And with that brief and non-descript revelation the ThreeofUs headed for the black archway on the far side of the room. Stepping into the darkness, he disappeared. Jerry stood briefly in the room for a moment, wondering if he too could simply pass thru the darkness into whatever waited on the other side. He stepped towards it slowly and walked into what seemed to be a black wall of darkness that was impermeable to him. Surprised, he turned to the other archway where he had entered. It looked identical to the one at the far end of the room. He moved toward it and instead of trying to walk thru, he touched the darkness and to his surprised his hand went thru. Stepping in slowly he exited the archway to the still waiting Sergeant Barclay.

"That was fast," Barclay said as Jerry approached, looking around and wondering how the Us had managed such a trick.

"What do you mean?" he asked Barclay when he reached his escort. "We were in there for a while."

"Sir, you were only gone a—"

Jerry shook his head, holding up a hand for quiet. How long he had been in the castle was no longer important. Not talking and staying alive were the only important things to be concerned about at the moment. "Let's go."

Troyconics Astrophysics — Columbus, Ohio, U.S.A.
Earth Date: October 25th, 2014 — 8:36pm EST

When Jerry and Sergeant Barclay made their way back through the Dimensional Mirror that evening, almost five hours after they had left, they were met by a room full of engineers, videographers and army personnel. The lab was bustling, and Jerry didn't like the way that scientists he didn't know were pouring over the specs of his creation, taking notes and examining his methods.

"Calvin, what's going on?" Jerry demanded as he walked down the gangplank.

"What are all these people doing here? Who are they? And what happened to the portal? I can't see thru it anymore?"

"Rajin made a suggestion Jerry that it may be a good idea to add a shield to the portal. He said you and he had worked on it years ago together and it only took a minor adjustment to add it," Calvin explained.

Rajin chimed in, "Jerry I simply made an adjustment to the light spectrum. It provides a pretty good shield to prevent an unwanted guest or fires."

Jerry gave him a look, the wound obviously not healed from the accident. "Either way," Rajin continued, "The slight change makes the light hit the portal more evenly."

"So it's now functioning or appearing more like specular reflection instead of diffuse reflection?" Jerry asked.

"Exactly! So now the portal *really* looks like a mirror to our eyes, even though it's still really see through,.. just not to us," Rajin exclaimed.

Calvin appeared to be confused by the explanation and hoped that it had distracted Jerry from any other inquiries. Jerry however simply moved on to other targets.

"Calvin who are all these people? We don't need them!!!" Jerry said frustrated.

"Jerry, leave it alone," Calvin told him, meeting him at the bottom of the ramp. "This comes directly from the President himself."

Jerry was no longer listening at that point. He was too busy shoving one of the army's engineers away from one of the power couplings. "Hey, don't touch that!" he yelled.

"Stand down, Jergensen!" Jerry heard from behind him and whirled around to see General Maddox standing there, a smug grin plastered across his face.

"This is your doing," Jerry hissed.

"Don't look at me," Maddox smiled. "I'm just the messenger."

"Right. So? What is this about?"

"Oh nothing," Maddox told him. "We just decided to stop horsing around with you and brought in our own engineers. The President wants additional Dimensional Mirrors built. He doesn't think it's wise to have just one. He shares my love of redundancy and preparation, you know."

"Really?" Jerry said. "And who's going to oversee the operation of these other Mirrors?"

"You will," Maddox offered, "if you act right. Otherwise, we'll find somebody else. Jergensen, don't act so surprised. The moment we bought this place out you knew where this was going. It's just happening a lot faster than you or I thought."

"You have no idea what you're dealing with," Jerry insisted.

"Maybe not," admitted Maddox, "but I will after your debrief. Meet me in the conference room tomorrow morning at 0700 hours. And be prepared for a full debrief. Sergeant Barclay!"

"Yes sir?" Barclay said, now out of his suit and walking over to them.

"You're with me," the General said as he walked away from Jerry and Calvin. Barclay followed and together they disappeared through the slowly opening blast doors.

Jerry whirled on his boss. "Calvin!"

"Jerry, this is above both of us. Even Maychoff. Leave it alone. They're running the place now and there's nothing any of us can do about it. How did the trip go?"

Sighing, Jerry began to take off his own suit. "Better than expected," he said. "There are people there...kind of."

"What do you mean? Are they aliens?"

"No. Or yes? I mean… they don't live on Earth so I suppose they are, but I think they used to be humans. They knew things… things about Maddox. And they saw us when we went in last time."

"What? How do you used to be human? I'm going to need a debrief before the debrief."

Jerry let out a chuckle. "Got time for dinner?"

Calvin opened his mouth to answer, but a familiar voice interrupted him before he could say anything. "I do," Maychoff said from behind them.

"Mr. Maychoff," Calvin said, surprised. "I thought you were in London."

"I was," the CEO said as he joined them. "I just got back. Having one's own jet has its advantages. So, what do you gentlemen have a taste for?"

"Wherever we go," Jerry said as he pulled off his boots, "let's make sure it has a private dining room."

Silcotti Waterfront Bistro — Westerville, Ohio, U.S.A.
Earth Date: October 25th, 2014 — 9:22pm EST

A large stone fireplace crackled with flames that warmed the private room of the bistro. Maychoff, Jerry and Calvin were enclosed in the elegant dining room, seated around a custom cherry wood table and enjoying an expensive bottle of red wine that Maychoff had picked out himself.

"So Candy takes me to church after dinner," Jerry was in the middle of telling them, "where this minister is teaching and he starts talking about words. How words run everything. And I'm sitting there and I start putting the puzzle pieces together."

"What pieces?" Maychoff asked, sipping from his glass.

"Atkins and Johnson were speaking negative words like 'hell' and 'damn.' Victor specifically said 'we're not even here for a day and we're dead already.'"

"What about Lilliana?" asked Calvin.

"Lilliana simply got caught in the rear with Atkins. He had her by the hand and she could not get free. I never heard her say anything. So then this guy comes out of the castle. He looks human—like a teenager—but he clearly knows things. He knows my name. He knows I have superiors. He says Sergeant Barclay's heart 'isn't right' and forbids him to come into their... he called it their sanctuary.

"So he invites me in," Jerry continued, "and I tell Barclay to stay outside. He takes me through the building and out onto a balcony, where he tells me all these things about words. I mean, it's like Déjà Vu, because basically everything Candy's minister said is what the ThreeofUs said...sorta."

"Three of who?" Calvin asked, confused.

"His name is the ThreeofUs. There are actually twelve of them."

"There are more?" Maychoff said, in awe.

"At least twelve. Who knows, there could be even more."

"What did they say?"

Jerry shook his head. "They others never spoke. Only the ThreeofUs. So he starts explaining to me how things work on the planet..." Calvin and Maychoff listen intently as Jerry shares *most* of the other details of his visit.

Troyconics Astrophysics — Columbus, Ohio, U.S.A.
Earth Date: October 26th, 2014 — 7:41am EST

The next morning, Jerry found himself once again seated around one of Troyconic's many conference tables in one of Troyconic's many conference rooms. Calvin, Maychoff, Barclay and General Maddox were also there, along with a couple of assistants. Jerry had just finished debriefing the General, telling him everything—or *almost* everything—that had happened on the planet Dabar.

"So is that it?" Maddox demanded. "You got some bozo with a name from *The Matrix* who tells you that three of your teammates died because they basically said the wrong words?"

"I told you the truth," Jerry said with a shrug.

"Yeah, and I have nothing but your word to take, seeing as how you left the only other member of your team outside while all of this transpired."

"It wasn't my call," Jerry said defensively. "He told me that Barclay couldn't come in."

"He who?" Maddox laughed. "Number man? Who is this guy anyway? And why is he giving orders?"

"It's their planet!" Jerry reminded everyone. "We don't have people popping out of mirrors and telling us what to do. I don't think that would go over too well."

"It's true, Sir," Barclay spoke up. "The teenager—or *whatever* it was—said I had to stay outside."

Maddox frowned. "And what would he have done if you had insisted? Did he have any weapons of any sort?"

"No, Sir," Barclay reported. "At least not that I could see."

"They appear to be very peaceful," Jerry reiterated. "And forcing your will upon others is not a great way to start a relationship. I don't believe I was in any danger at any time."

"Are you sure you're telling us the whole story, Jergensen?" Maddox grumbled, his stare all but boring holes in Jerry's own eyes. "You didn't happen to meet up with number man on one of your previous trips, did you?"

"No, I didn't," Jerry said, insulted by the implication. "When I went thru the first time by myself, I only stepped inside the portal briefly."

Yet the General did not seem convinced. "I find it very interesting," he said, "that he doesn't know you, but when you show up you're the only one that he'll talk to…" Maddox held Jerry's gaze for a moment, sizing him up, then turned to Calvin. "Here's what I need to happen, Madison. Since it is pretty much assumed that there is no real threat, I want to take two teams back through the mirror, including myself. And since Mr. Jergensen is now on a first name basis with Number Man, he can make an introduction and we can find out for ourselves what we need to know."

"What do you mean there's no threat?" Jerry cut in, before Calvin could respond to the General. "There was no threat because I made sure rules were followed to prevent a threat."

"Good!" Maddox shot back. "Then you'll have no trouble sharing those rules with me so that I can share them with my team."

"Sure General. Would like to have those rules right now?"

"Why that would be awfully kind of you," Maddox said with a condescending smile.

"Well good," Jerry said, his face a slab of weighted seriousness. "Are you ready? Here it is: *shut up*!!!"

The General's face flushed with anger, and seemed to bristle through the dark green of his decorated uniform. "Who are you talking to?" he bellowed.

"You," Jerry said, not backing down. "You asked for the rules, and there they are. When we go in, keep your mouth shut."

Barclay nodded to the General. "That's what he told me to do as well, Sir."

"It's a simple rule," Jerry continued. "But can you and your men follow it? Because believe me, if you or your men decide they know more than I do about this, then don't forget to add one more thing to your list of supplies…"

"And what's that?" Maddox growled.

"Body bags!"

Jerry realized then, that his fists were shaking. He refused to let any more lives be extinguished because nobody would listen to him. Pushing his chair over, Jerry shot up out of his seat and stormed out of the room. The General looked as if he were about to explode. "Excuse us," Calvin said as he ran after Jerry.

"Sergeant Barclay," Maddox said through gritted teeth.

"Yes sir?"

"Assemble our teams and make sure they're prepped and ready to go."

"Yes sir."

Jergensen Residence — Westerville, Ohio, U.S.A.
Earth Date: October 26th, 2014 — 9:01am EST

Setting down her steaming mug of coffee, Candy switched on the television, getting ready to start her morning routine. It was laundry day, and on laundry day she liked to fold clothes while she drank her morning coffee and watched the news.

"*Good morning,*" the UNN news anchor said through the TV, "*and welcome to United News Network. Today's top story is Patricia Slater, the senator from Colorado, who is running for President in the upcoming election has taken a massive lead in the latest polls. Those polls indicate Slater to be in the lead against Drake Duffy by a whopping fourteen percent. The Senator had this to say:*"

Candy folded a T-shirt as the TV cut to footage of a press conference. A woman in a crisp pants suit stood behind a podium, a bouquet of microphones just beneath her face. A graphic labeled the woman as Senator Patricia Slater.

"*I think it's just an indication that America is ready for a much needed change,*" she was saying into the mics, "*and whether that change comes from a male or a female, the American people don't really care at this point. They're beyond that.*"

"*The Senator is making her way through the final leg of the campaign trail,*" the news anchor took over again, "*shoring up the Midwest, specifically the key state of Ohio. A smart move on her part, as elections are only four weeks away.*"

Candy picked up the remote control and beginning flipping through channels. She liked the news for its news, not for its political analysis. As she surfed through the stations, she called upstairs. "Lexi! Are you ready to go?"

"Five minutes, Mom!" her daughter yelled back from her bedroom.

"Okay! Just remember, I have somewhere to be too!"

"Mom, really? Okay!"

Sighing, Candy finally settled on a network. It was the Christian

Community Network broadcast, and Candy was both pleased and surprised to see the guest-speaker who had come to her church on the screen. He was ministering to the audience again—a bigger one than her congregation—and he paced the floor of the venue as he spoke.

"*In Proverbs chapter 18 verse 21,*" he said, "*it reads… The tongue can bring death or life; those who love to talk will reap the consequence. One of the most difficult things to get people to understand and this includes believers and non-believer, is that their words actually do things!*"

"*Words are containers and you can choose at any point and time to speak words of life or words of death. Now many people unfortunately choose to speak words of death not only over themselves but their loved ones. Things like I'm dying for a piece of cake or that just scared me to death.*"

Just then the sound of the television was overtaken by the rumbling footsteps of Alexandria charging down the stairs. "Okay, Mom. Let's go!"

"Just a sec," said Candy, holding up a hand, enthralled by the teaching.

"Mom, really? After you were rushing me?"

"Just let me record this program for your dad. I think he'd want to see it."

"Since when is Dad into church?"

Candy smiled. "You'd be surprised what your Dad is into." She thumbed the remote until she was sure the DVR was saving the show, then turned off the TV and finished her coffee. Then together, the two Jergensen women grabbed their purses and headed out the door.

There was a *beep* from Candy's phone as they approached the car and when she checked the screen she found a text message from Jerry: *Leaving the planet again soon. Call you when I get back. Love you.*

Candy let out a loving sigh and a slight smile. "He does try…"

Chapter 6

The third Troyconics insertion team that would make contact with Dabar was assembled in Jerry's lab. They were military mostly, and they stood at attention, listening to every word Jerry had to say. Not because they were interested, but because their General had ordered them to. Jerry, however, didn't care why they were listening, as long as what he had to say sank in.

"I am not your superior or leader," he told them. "That would be General Maddox. However, if each and every one of you values your life and would like to return from this mission, you should listen to me as if this was a direct order from General Maddox."

There was an audible sigh from behind him as General Maddox rolled his eyes, but Jerry kept going. "The planet where we are going to visit has a unique function. Everything on the planet is controlled, happens, or takes place by words. Words that we consider normal or harmless can kill you on Dabar. So please, do whatever it takes to restrain your self from words such as damn, hell, death, dying… Don't say 'we're going to die; or say anything that you would not desire to actually *take place*."

Jerry heard a quiet snicker from somewhere in the amassed group and he wheeled around, searching for who had made the sound. "Excuse me, Private, did I say something funny? General, perhaps he needs to be replaced."

"He's in, Jergensen," Maddox shot him down.

Jerry stared at the young man who had laughed. He was watching Jerry with a condescending smile. As if Jerry were some kind of stand up act in a comedy club. "What's your name?" he asked.

The Private looked over Jerry's shoulder to his General, who nodded. "Jeremy Tanner," the Private finally said.

"Well, Mr. Tanner," Jerry scowled at him. "I'd keep my smirks or other comments to myself. Anyone else got a joke? Cause I can guarantee you and your family won't find it funny if a few hours from now you're in a body bag!"

That seemed to wipe the smile off the Private's face. In fact, the entire room seemed to tighten in a collective inhale at Jerry's warning. He nodded. "Good. No laughs on that. The best rule to use once we go through those mirrors is to keep your mouth *shut*. Period."

"So how are we supposed to communicate?" another team member— this one wearing a Corporal's uniform—asked.

"Hand gestures," Jerry said. "You guys have those, don't you?"

"Man, this is some bullsh—" the Corporal started.

"You see!" Jerry cuts him off while getting in the man's face. "That's what I'm talking about! One little comment that you so casually ramble off here can get you killed there. Start practicing shutting up like right now. Because I kid you not, *this is real*. We head out in one hour. General?"

Maddox cleared his throat and stepped forward beside Jerry. "You heard the man. Practice your hand gestures and practice shutting up. I'm coming back with everybody I left with…alive. Is that clear?"

"Sir, yes sir!" the two dozen uniformed men shouted at once, the sound reverberating through the laboratory.

"Dismissed!" the General shouted, and the men dispersed, each of them going off to prep for the insertion.

Dimensional Mirror Insertion Point — Dabar
Date & Time Unknown

As if from nothing, the portal opened and Jerry stepped through. General Maddox was close behind, and soon, the entire team was through. They stood in the alien countryside, marveling at the lush valley that cradled them. Jerry looked around, a bit troubled. Everything seemed a little different than on his last trip. He looked over at the closest six-second box and stared at the dark screen, watching it closely.

Maddox gave instructions to his two teams. "Major Donaldson," he said, "you and your team remain here. The Alpha team will go with me."

The suited Major nodded, but Jerry barely noticed still watching the six-second box like a hawk. Maddox walked up to him. "What is it, Jergensen? What's the matter?"

Jerry shook his head. "Nothing. Let's all keep quiet."

Together, they made their way up the hill again and up to the Sanctuary of the Us. Jerry stayed up front, the default ambassador. As the castle came into view the ThreeofUs was already standing outside when they approached the castle.

"Good evening, Jerry. I told you I would see you soon. And you've brought company."

"This is General Mad—"

"General Maddox, yes," ThreeofUs said. "And Lieutenant Colonel Brian Andrews. And Field Specialist Tamyra Anns. And who could forget Sergeant Barclay."

Unfolding his hands from behind his back, ThreeofUs produced a pair of sealed envelopes. With one in each hand, he handed them to Jerry and Tamyra respectively. "Jerry. Tamyra. These are for you. Please, do not open them until after you leave."

"Sure," Jerry said, puzzled. "No problem." He tucked the note into his pocket.

Tamyra did the same with hers. "Thank you?"

"Thank *you*," ThreeofUs replied.

"It's, uh, an honor to meet you, um, Three," General Maddox tried clumsily.

The ThreeofUs turned to him, a blank expression in his teenage face. "Lying is distasteful here. As is the misuse and the calling a being something other than their name. However, because we have dispensed with pride, We will overlook your disrespect. Even if you prefer to continue referring to Us as 'Number Man.'"

The General turned red in the face and turned to Jerry, who couldn't help but smirk. The ThreeofUs continued, "Do not be shocked. When you come to our world, your words are known to Us. As we have nothing to hide, nothing can be hidden from us. However if you prefer, we will hide your words."

"What exactly do you mean?" Maddox demanded.

"Simply that if you do not desire to be made aware that We are aware, We will not make you aware of what We are aware of. But this not to say that We are not aware of it."

The ThreeofUs stared unemotionally at the General, who looked to Jerry again, confused. "What did he just say?"

Ignoring the General, Jerry turned, and spoke directly to ThreeofUs. "Can we speak with Us?" he asked politely.

"Jerry," ThreeofUs smiled, "your quickness to learn and respect of others is what has attracted Us to you. We have been looking forward to your next visit."

"General Maddox would like to be a part of our conversation," he told the ThreeofUs. "As well as the Lieutenant and Field Specialist. Is that acceptable?"

The ThreeofUs walked over to the General, looking in and through him. "Although we see the same selfish motives and abuse potential in the General, which caused us to decline the Sergeant's entrance, we will allow it because you asked, Jerry. Sergeant Barclay, however, must still remain outside."

They followed ThreeofUs into the sanctuary. The inside was still a jumble of backward and upside down physics, but it had changed since

Jerry's last visit. The furniture had changed; even the size and layout of the castle had been altered on the inside with no apparent change to the outside.

"We made some modifications to the asylum to accommodate you and your comrades," ThreeofUs explained. ThreeofUs motioned to the empty chairs. "Please be seated."

"Forgive my ignorance," Maddox spoke up, "but who are"We?"

"They speak of all their selves as one, General," Jerry tried to explain. "You'll get used to it."

"Perhaps," ThreeofUs corrected. "Sometimes the selfish constantly depart, but never arrive."

The General looked offended by this remark and opened his mouth to say something, but Jerry shook his head in warning. "General…"

Maddox seemed to swallow his anger and tried again: "I apologize if I have offended you in any way. We would like to learn as much as we can about your world and how it functions and operates. We would also like to discover any and all opposition that you—"

"*Us!*" the ThreeofUs boomed, without actually raising his voice.

"Excuse me?"

"You stated 'you.' The correct reference is 'Us.'"

"My apologies," General Maddox grumbled, frustrated. "Is there any opposition the 'Us' would have to us allowing our people to come here and populate designated areas of your planet?"

Jerry's head whipped around to face the General. "Wait a minute. We never discussed this! Are you talking general population?"

"Jergensen, this is above your pay grade. Stand down."

"Who is this coming from?" Jerry wanted to know.

It was the ThreeofUs that assuaged his concern. "Jerry," he said calmly. "Let there be no strife among us. We told you on your last visit that We realized it was you and your people's intent to continue coming here, and We welcome it. We are not a selfish people."

Then the ThreeofUs turned to Maddox. "However General, although We have not and do not intend to control the planet with a totalitarian rule, please be aware as Us, We are its overseers and will continue to make sure that decisions are made in line with its best interest. Mutual respect is an expectation."

"Of course," Maddox said. "It is our desire and intent to work *with* you—I mean Us—and learn from, uh… Us as well."

"Really?" ThreeofUs asked. "To what end?"

"Hopefully toward the end of a mutually beneficial relationship. I am sure that there are things that we can offer y—Us as well. So. Is there an area you would designate for us to populate and use?"

The ThreeofUs nodded. "Initially there is roughly an eight thousand square mile section to your north. You may begin with that. If handled properly, the entire planet is before you *with the exception of what belongs to Us*. However, in terms of teaching your people how to live here on Dabar, your people will require much training as things are done very differently here."

"What do you suggest?"

"I will train Jerry," the ThreeofUs allowed. "And eleven other civilians."

The General didn't seem to like that idea. "We prefer to have our own people be trained. It tends to make things a bit easier."

"Easier for whom, General?" ThreeofUs asked, to which Maddox had no response. "You must work on promoting selfish agendas. I will allow one person from your team to join the trainees."

"Alright then," Maddox said. "I volunteer myself."

ThreeofUs shook his head. "You have already disqualified yourself by way of your actions and attitude. And I must say, I am taken aback by your attitude. During my time on Earth, it was customary that one who was chosen to be in authority had to show superior skill at being *under* authority in military service. It is a flaw of your people to believe that all rules and laws need be changed to match the time. And forgive me, General. Here, in most cases, the truth does not offend but enlighten. Jerry is a good judge of character. We trust his choices."

"I'm still working on it," Jerry said modestly."

"You have respect and integrity," ThreeofUs went on, "and that is enough. Find eleven civilians and bring them before Us, preferably people of faith. This will make the training and transition much faster and easier. If at all possible, children are the best candidates."

"Why?" Jerry asked. "Why would you ask for a child knowing that they could possibly harm themselves by what they say?"

"I will answer the second question first," ThreeofUs told them. "All children age twelve and under are not held responsible for their negative speech. They are trained in proper speech until age twelve, and at age thirteen their words then manifest their creative power in both positive and negative ways. Until that time their negative speech is not counted against them. As to your first question: 'why?' It is quite simple. *Children have less things to unlearn.*"

Then ThreeofUs turned to Tamyra. "Are you with the military? Or are you a civilian?"

"I work as a civilian contractor," Tamyra stammered.

"She can be one of Us," ThreeofUs said.

Jerry looked at the ThreeofUs curiously. "What do you mean?"

"She can be one of the eleven, as you are one of the twelve." Then the ThreeofUs stood up, graceful as ever. "We must go now, Jerry. General, Lieutenant, Field Specialist… It was a pleasure *seeing* you. Jerry, you know the way out?"

"Yes," Jerry said.

The ThreeofUs stepped toward the darkened entrance to the remainder of the sanctuary, but stopped. "Is there anything that you would like to ask Us?" without turning around.

"Yes," Jerry said. "Will you allow Sergeant Barclay to be the military trainee?"

"Yes," ThreeofUs surprisingly told him. "We will allow it. We denied him entrance because of his disposition at the time, and as a test and lesson for later. Is there anything else?"

"I'm sure questions will come up during the training," Jerry said. "I'll wait until then. Thank you. General?"

"No. No questions here. You've been more than accommodating."

The ThreeofUs finally turned back to them, staring at the General. "Yes. *We* have! We can only hope that the same courtesies are reciprocated back to Us." When the statement was met with silence, the ThreeofUs continued. "There is one more stipulation that we need to be perfectly clear and agreed upon."

"What's that?" Maddox asked.

"Once your people begin to come in abundance, they will desire to bring their machines and their ways to Dabar. Everything that is created on Dabar *must* be done so by words. *Anything* created in another way, such as by machines or your normal way of manufacturing, must be restricted to an area on the planet called Lodebar."

"Okay," the General said. "May I ask why?"

"You may," ThreeofUs said.

"Oh," Maddox said after a brief moment of silence. "Umm, well, why then?"

"Because anything created on Dabar that is not created by words will not last. It will cease to exist at some point. It would be unfruitful for a building to cease to exist while there are hundreds of people still inside it. Or a house, or a car. Do you not agree?"

"That can happen?" Jerry asked, horrified.

The ThreeofUs stared at Jerry, making deep eye contact. "And will!" he said. "Someone will disobey."

And with that, the ThreeofUs disappeared through the dark doorway and further into the castle. Jerry led the other three team members back through the entrance. They gathered their team and headed back to the insertion point where the others were still waiting in unusual silence.

There was something wrong and Jerry knew it. As they got closer, they could see a few soldiers crowded around something. There, in the center of the cluster, was Corporal Leonard Shank. He was lying on the ground. Dead.

The General rushed over and knelt beside the deceased man. "What the—"

"*Shut your mouth!*" Jerry yelled. "Pick him up, let's go!"

The General nodded and relayed the order to the rest of the men. Together, they picked up Shanks body and began carrying him back toward the mirror. Back toward home.

Troyconics Astrophysics —Columbus, Ohio, U.S.A.
Earth Date: October 26th, 2014 — 4:52pm EST

"No! No! No!" Calvin yelled as the team stepped back through the mirror portal. Four men were clustered together, struggling to carry something heavy between them. As soon as they reached the bottom of the gangplank, Calvin knew it was a body. "Jerry! What in the world happened?"

"Blame the General," Jerry yelled as he angrily stripped off his gear. "I told you this was the main reason I didn't want to take anyone else through besides myself!"

"Jerry," Calvin said, grabbing his shoulder, trying to get a straight answer. "What happened?"

"You know what happened," Jerry said, shoving Calvin off of him. "Someone couldn't keep their mouth shut."

Meanwhile, a livid General Maddox was stomping down the gangplank, looking for the man he had left in charge. "Donaldson! What the hell happened?"

"Sir, it was Jamison," Major Donaldson tried to explain, terrified. "A little while after you left he slipped Shank this note." He handed a scrap of paper over to the general.

"I dare you to say 'I'm dead,'" the General read out loud. "Pussy…"

Donaldson nodded solemnly. "They giggled about it for a while. Then Jamison took the note back and added that last part. Called Shank a pussy. Then he handed it back. I ordered them to be quiet but Shank got pissed, and he stood up and said, 'okay I'm dead!' There was a beeping sound and one of those little parking meter things lit up with the number six. Shank got freaked out. He grabbed Jamison and said 'what do I do? What do I do? Oh crap, I'm gonna die!' Then the box started counting down. I motioned to him to be quiet and everyone else did the same, but I didn't know what else to do except for what Mr. Jergensen said. To keep our mouths shut. When the box got to zero, Shank fell on the ground. We tried to give him CPR, but he didn't respond. So we… We just sat there in silence waiting for you to return."

Maddox processed Major Donaldson's story. Then he whirled around, eyes like daggers slicing through the scared expression of Lieutenant Jamison. "You dumbass," Maddox growled. "I'm gonna bring you up on charges of insubordination and disobeying a direct order. And if I can, you'll be brought up on involuntary manslaughter." He punched the soldier in the face. "Do you know what you've done?"

"Sir!" Jamison cried, falling to one knee. "I am so sorry, Sir. It was a joke! I never thought anything would happen—"

General Maddox swung his arm in a powerful arc, the back of his hand connecting with the other side of Jamison's face as he then went down on both knees. "Did you hear anything that you were told in the prep meeting?"

Tears were streaming down Jamison's face now. "Yes sir, I did... I just... I just didn't believe it, Sir..."

Jerry walked up to them, getting in the General's face. "And you think you can bring regular civilians to this place?"

Maddox had nothing to say and, after a short staring contest, Jerry stormed out of the lab with a huff. "Everyone prep for a complete debrief!" the General finally barked at the rest of the team before leaving himself.

The men and women dispersed, stripping their gear and attending to Shank's corpse. Tamyra Ann, the Field Specialist, took herself into a quiet corner of the lab and slipped the envelope ThreeofUs had given her out of her pocket.

When you return for training, the note inside read, *please bring LEGO set item #10225. Thank you. — All 3OfUs*

"LEGO?" Tamyra murmured under her breath, wondering what the strange request could possibly mean.

5:08pm EST

Moments later, as Calvin filled out paperwork regarding the death of Corporal Shank, Jerry burst into his office. He had stripped out of his gear and changed back into his regular clothes, but his temper was still ablaze. "Calvin," he said, shutting the door behind him. "This is getting out of hand. I'm beginning to question whether or not I should have ever exposed the mirror in the first place."

"No you're not, Jerry," Calvin reminded him. "You're just upset that you have no control over it now. This is still your passion, your baby and you know it."

"Yeah, I guess you're right…" Jerry sighed, plopping down in one of the chairs across from Calvin's desk. He was exhausted. "By the way, when we got on the other planet and spoke with the ThreeofUs, General Maddox asked for permission to populate designated places on the planet with general population. Civilians…"

Calvin kept his head down, scribbling out words on the paperwork in front of him. Jerry cocked his head in shock, staring at him. "Did you know about this?"

"Jerry, I knew," Calvin said, taking a break from the paperwork and leaning back in his own seat. "I couldn't say anything because I was under a gag order. They thought you would go ballistic and not take them back to the planet. Jerry, the President is pushing this. He sees it as an opportunity."

"The President may not even be in office in a few weeks," Jerry protested.

"But he is now!" Calvin shot him down. "Jerry, you're so caught up in the middle of this, I don't think the magnitude of the entire thing has hit you. And because of the deaths, you're only focused on the negative possibilities instead of the positive potentials. Imagine it: people creating all types of things from literally *nothing*."

"It's not 'nothing' Calvin…" Jerry told him, shaking his head.

"Look, I don't pretend to understand it all, but it seems if a person can learn how to talk and how not to talk, life there can be a lot easier and more fulfilling than it is here. Do you see what I'm saying, Jerry?"

Jerry thought about it. He let Calvin's words wash over him and sink in. All at once, he saw everything from an entirely new perspective. "You know… For the first time, I think I do…"

"Jerry, this thing can be bigger than the internet. I think someone else is seeing dollar signs also, and that's why they're pushing this thing so fast."

"I see it, Calvin," Jerry said, nodding. "But it's smart and dangerous all at the same time. If that's what's being done, someone's making a decision that some people are expendable. Calvin, they have to know, *you* have to know that *people are going to die!*"

Now it was Calvin's turn to sigh. "And Jerry, that's where you come in. Now obviously, you and the ThreeofUs have some type of connection. Use it to find out everything you can from him so you can teach people how to master life on the planet and stay alive. Are you going to have some people who won't listen? Of course! But Jerry it's no different than what happens here. People die every day right here on Earth taking chances. That's their choice. It's a chance and a choice, Jerry. And just like here on Earth people have the right to make their own decisions." Jerry begins to see the light of what Calvin is saying and he begins to change.

Troyconics Astrophysics — Columbus, Ohio, U.S.A.
Earth Date: November 10th, 2014 — 3:27pm EST

Having only been based at Troyconics for slightly under a month, General Maddox had already decorated his office in the manner he liked. It was minimalist and formal, furnished with functional desks and chairs and stylized with a few pieces of his favorite war memorabilia.

Content in his large office, the General was currently reading reports from the other building sites and cross-referencing them with each other when there was a *knock* at the door. "Come in!" he called, glancing up towards the door as it swung open. It was Jerry. "Jergensen. To what do I owe the honor?"

Jerry walked over and dropped a stack of files on top of the reports the General was reading. Maddox picked them up and flipped through them. "So, you finally got the trainee list finished. What took you so long?"

"It's not something you want to rush through," Jerry told him. "People need to be evaluated and some people we approached weren't open to it at all, which meant we had to identify others."

"Over half the list are Troyconics employees. Some of them kid's. I take it that's deliberate?"

Jerry nodded. "We pulled some viable candidates from here that are in roles that could be performed from the other planet. A parent is not going to send their child to another planet without them being there. They asked for children."

"Doesn't that strike you as a little weird?" the General asked.

"Not at all," Jerry exclaimed. "Kids are open to learning new things. It makes perfect sense."

"Have you talked to Tamyra?"

"She's already on the team. Is something wrong?"

"She's been having second thoughts," the General informed him. "Would you mind talking to her?"

"You mean talking her into it?" Jerry said with a raised eyebrow.

Maddox shrugged. "Whatever works."

"Did you see the other file?"

The General moved the list of trainees to the back of the pile and thumbed through the other pages. "Requested items. What is this?"

"It's called a chastity muzzle," Jerry explained.

"Looks like the mask from *The Silence of the Lambs*," Maddox said.

"It's kind of like that, with some modifications. As you can see, the insert prevents the tongue from moving and forming words, while the air hole allows people to continue to breath. We'll need to work this into the overall budget of the project. Every person who comes to the planet and starts training will need one."

"Says who?" Maddox asked.

"Says me," Jerry told him. "General, I've already faced the truth that more people are going to die, and I know you know this. I take it as my personal responsibility to make sure those numbers are as low as humanly possible."

Maddox looked from Jerry, down to the design for the chastity muzzle, then back to Jerry again. "Okay, Jergensen. I'll have it worked into the budget."

"We'll need it in a variety of sizes for adults and children."

"Understandable. Should take about two weeks to mass produce."

"Well we're going to need twenty-four right away for our return trip to Dabar," Jerry pointed out.

"I'll see what I can do," the General said. "When will your team be ready to go back in?"

"Most are ready right now," Jerry reported. "I'll talk to Tamyra as soon as I can."

"Good. Thank you, Jergensen."

"Sure thing," Jerry said, and walked out of the office.

Ming Flower Chinese Restaurant — Worthington, Ohio, U.S.A.
Earth Date: November 11th, 2014 — 12:41pm EST

Nestled into the corner of a small strip mall right outside Columbus, the restaurant was quiet and slow. The cloth napkins and glass lazy Susan's that adorned each table were pristine, and the speakers were softly pumping classical Mandarin music into the dining room. In a booth by the window, the only two patrons in the place talked quietly with each other.

"Lisa," said Sergeant Barclay to the woman who sat across from him. "I'm glad you could join me on such late notice."

Lisa Hansley shifted in her seat, unsure of why she had been invited or why she had come. "Well," she said, "you said it was important. But, it does seem kind of weird… I mean, seeing how we never really talk at the office."

"I apologize for any inconvenience. But General Maddox asked me to speak with you."

"General Maddox," Lisa repeated. "But he doesn't talk to me either, unless he needs something."

"Well Lisa," said Barclay, trying to sooth her uncertainty. "I'm sure it would seem even more strange had the General asked you to lunch himself. Just think of me as a buffer for the General."

"A buffer," she said, nodding. "What exactly is this all about?"

"You were recently selected for a new project. Correct?"

"Yeah, but we're not supposed to talk about it with anyone," she told him.

"Well, Lisa, the General is overseeing that project and saw an opportunity for you to get a promotion and raise all in one fell swoop."

Lisa leaned back against the booth, looking around to make sure she wasn't being watched. Or perhaps she was on one of those hidden camera shows. "This is getting really weird. All this sudden interest in me."

"Listen Lisa," Barclay said. "The General would like to make you a sort of… interplanetary liaison."

"What do I have to do?"

"Just what you've already signed up to do," he told her, "be a trainee. However, the General would like you to forward a report to him of all your activities and interactions with all personnel involved, and maintain that report on a daily basis. Then, when you come back for debriefs or leaves, just simply forward that report to me. If you're interested, it will take you from your current salary of $45,000, to $60,000."

"Sixty?" she said, her eyes almost popping out of her skull. "Did you say *sixty thousand dollars?*"

Barclay smiled as he watched Lisa try to keep her cool. A second later, a waitress walked over, holding a steaming ceramic pot. "More tea?" she offered.

Lisa seemed frozen in her seat, thinking about the potential raise. "No thank you," Barclay answered for both of them, then pausing so the waitress could walk away. When she was gone, he turned back to Lisa. "If you could keep your voice down... Yes. Sixty thousand."

Lisa nodded, almost forgetting to breath. "I'm in," she said.

Barclay grinned. "Great. We'll either provide you with a laptop if the planet is able to provide power, or a small recording device to record your reports daily. One other thing, Lisa. No one can know what you're up to. Not anyone."

"Not even Mr. Jergensen?" Lisa asked.

"Lisa, you've been here long enough to know that when money is involved, people can get jealous. We don't want any strife on the team. So not Jergensen, or even Madison can know. You report directly to the General. So can you do it?"

"Yes," Lisa said, nodding emphatically.

"Great. That increase will be effective on your next paycheck. It will come from a separate check due to the fact it's being paid from a different department. If you need to reach me, here's my number." Barclay took out a business card and slid it across the table to Lisa. When she reached for it, he put his hand on top of hers and looked her in the eyes. "Lisa, thanks. This will really help us achieve our goals for the project. Don't forget: *do not* share this with anyone *inside* or *outside* of the office. That's part of the deal."

Lisa nodded. "I got it."

"Great," Barclay said, then raised his hand and gestured for the waitress. "Can we have the check please?"

Troyconics Astrophysics —Columbus, Ohio, U.S.A.
Earth Date: November 11th, 2014 — 1:45pm EST.

The break room was the busiest Jerry had ever seen. Not that he spent much time there—most of his days were spent down in his lab—but it still seemed more lively than ever. Employees clustered around tables, talking and laughing. In fact, the only person who was seated alone was Tamyra Ann, the Field Specialist. Her only companion was the new flat screen that Troyconics had installed for its workers, and now it even had sound. She watched it intently as it spewed out the day's latest news.

"And still topping headlines," said the news anchor from UNN, *"is Senator Patricia Slater's landslide victory over Drake Duffy to become the next President of the United States. President-elect Slater was all smiles as she visited, once again, key states that were critical to her overwhelming victory, thanking them personally for their support. The Senator remarked that she would be taking a much needed rest after the campaign trail, and is headed to Las Vegas for a much needed vacation."*

Jerry made his way through the crowded break room and caught Tamyra's attention. "Is this seat taken?" he asked, gesturing to the empty chair next to her.

Looking up to find Jerry there, Tamyra smiled. "Hi Jerry. This is the first time I've ever seen you in the break room ever."

"Yeah, I normally pack a lunch or order out. I'm just anti-social like that."

"No you're not," Tamyra laughed.

Nodding toward the TV, Jerry said, "So, we're going to have our first female President. You excited? One big step for women, one giant leap for womankind?"

"Corny, Jerry. Really corny."

He shrugged. "I tried."

"Actually," Tamyra said, waving the TV away, "I wasn't even listening. I was in my own world."

"You just got back from another world," Jerry pointed out and they both chuckled.

"Yeah, and I'm debating going back."

"Anything I can help with?"

"No, just have some personal things going on. They've been kind of distracting."

"Well, there's nothing like going a few thousand miles away to take your mind off things."

"Yeah," she said, "but will it make it better or worse? If I go, it's not like I can communicate back and forth, can I?"

"We're working on that," Jerry told her. "Is it a relationship?"

"Yes," she admitted, almost embarrassed.

"Just tell them you have to go away for work for a while. You'll be inaccessible and you'll call as soon as you get back. Pulling away sometimes can give you…clarity."

Tamyra nodded, conflicted. "Yeah, I just think that too much clarity may mean that I pull completely out of it."

"Then you already know what you need to do," Jerry said simply and got up from his seat. "I hope you decide to come. You'll be a great asset."

"Thanks Jerry," she said, looking up at him.

"No. Thank you."

They shared a brief moment of friendly eye contact, before Tamyra's attention was pulled away. She was suddenly looking past Jerry, over his shoulder to the break room door. Calvin had just walked in.

"Hi Tamyra," he said, as he walked up to Jerry and her. "How are you?"

"Better," she told him."

"So are you going to be joining Jerry and the team for the next big trip?"

Now it was her turn to shrug. "I'm still not sure, but I am a lot closer now to making a decision than I was a few moments ago." She smiled at Jerry and got up from the table. "Thanks again Jerry."

"I'll see you later," Jerry told her. Then, once she had gone, Jerry pulled Calvin down into her old seat and they hunched over the table together. "What's the matter?" he asked.

"Four people on the list have backed out," Calvin informed him. "I knew this could happen. Asking regular everyday people to go to another planet is…not normal."

"Who backed out?"

"Who isn't important right now. I need you to start working on four or five more potential candidates so we don't get delayed again. General Maddox has been extremely patient, but I don't think it's going to last much longer. You got anyone in mind?"

"Actually," Jerry said. "I do."

Jergensen Residence — Westerville, Ohio, U.S.A.
Earth Date: November 11th, 2014 — 7:34pm EST

The house was quiet when Jerry returned home. There was no sign of his wife or daughter, but the melody of Canon in D Minor could be heard softly coming from the in house speaker system. Quietly, Jerry climbed the steps and made his way down the hall to he and Candy's bedroom door. It was ajar, and inside he could see his wife sitting on the bed, painting her toenails as she hummed along to the classical music.

"Hello," he said.

Candy jumped, startled, and the bottle of nail polish that she was using tumbled over, spilling onto the floor. Candy didn't seem to notice; she was too busy being surprised. "Jerry! What are you doing home?"

"How soon we forget," Jerry smiled. "I live here."

"Only enough to be a temporary resident," she said.

He climbed onto the bed and crawled across the mattress to her, planting a kiss on her back. "I'll help you clean up."

"Jerry, what's the matter? Whenever you offer to do something on your own, something is up."

Taking a deep breath, Jerry sat up, thinking of what to say. After a long pause, he finally opened his mouth. "I would like you to pray about something. I'm asking you to pray because if I ask you to do it outright you'll say no."

"How do you know?"

"Because I know," he told her. "*I* would say no."

"What is it?" she asked, turning to face him, concerned now.

"I want you and Lexi to come with me to the other planet with me for a few weeks."

Suddenly her eyes went wide. "*No way!*" she said, getting up off the bed and retreating into the bathroom. She came back with a washcloth and bent down, cleaning up the spilled polish.

"I said pray," Jerry reiterated. "Please."

"Jerry, I don't think I have to pray about this one. God put us on *this* planet and I think we're supposed to stay here. Now, if you want to go through your little Stargate fine, that's your job. But don't expect me—and especially Lexi—to join you."

Just then Lexi appeared in the doorway. "Join you for what?" she asked. Jerry and Candy looked at each other, not sure what to say. "You guys are weirding me out," their daughter said. "What?"

And so Jerry had no choice but to invite her into the conversation as well, and explain everything. He told her all that had happened to him in the last few months, and all that was about to happen. When he was finished, his daughter's face was a blank expression of disbelief.

"Sooo," she finally managed to say. "You really have been hanging out with E.T.?"

"Yeah," Jerry nodded.

"And now you want me and Mom to go with?"

"Yeah."

"Huh," Alexandria said. "How about *no*!"

Jerry sighed. "Honey, please. At least think about it. It's an opportunity of a lifetime. You'll get to do things you could never do here."

"Like what?" she wanted to know.

"Things…" Jerry stammered. "I can't explain it Lexi, you have to see it for yourself."

"Dad!" she yelled back at him. "School just got out. I had all kinds of plans to hang out with my friends and to get a summer job. Being an astronaut was not the job I had in mind."

"Lexi, there are people who would jump at the chance, but they have to wait. You can be one of the first. Please, just tell me you'll think about it? But I need to know in a few days."

"Isn't it dangerous?" Lexi asked.

"Not for us it won't be," Jerry assured her. "I know things. I know people."

"Who, E.T.?" she asked sarcastically.

"Kind of. Just think about it. Both of you."

Candy sighed from the chair she was sitting in. "Okay."

Jerry turned to Lexi, who sighed in the exact same fashion as her mother. "Alright Dad…"

"Okay," Jerry said, relieved and pleased that he got them to consider this far. "One more thing. *Do not tell anyone about this.* Lexi, I'm serious. I could lose my job."

"Alright already, Dad."

Jerry wrapped his arms around her and squeezed, pulling her in for a bear hug. "I love you, Lexi," he told her.

"Really Dad," she said before pulling away and disappearing downstairs.

"Lexi!" Jerry called after her.

"Yes Dad?" she called back to him from the staircase.

"Please don't call and tell Kendra—"

"Dad, I'm not!" she yelled to him. Though he couldn't see from the bedroom, Alexandria not only had her phone in her hand, but she already had Kendra's number up on the screen.

Troyconics Astrophysics — Columbus, Ohio, U.S.A.
Earth Date: November 12th, 2014 — 9:16am EST

The next morning, Jerry found himself back in General Maddox's office, reporting on the status of the next insertion. "So unfortunately General," he started, "four people have backed out. So we're in the process of filling those spots."

"Jergensen," Maddox said, and Jerry could tell from the grumpy tone of his voice that he had not yet had his morning coffee, "we've already been delayed for a nearly a month now. If you would have taken my team members we would have already been on that planet."

"Sir, you know that was not my decision," Jerry said defensively.

"So. Who have you got as backups?" Maddox wanted to know.

Jerry produced four manila file folders and placed them on the General's desk. He picked them up, opening each one in turn and inspecting the contents. When he got to the last two folders, he looked up at Jerry with a sneer. "Jergensen, you are a piece of work. I've always accused you of egotism. Now I can add another thing to the list: *nepotism!*"

"Sir, I—"

"Your wife and daughter? What is it? You can't go a few weeks without getting you some?"

Jerry could feel the blood rush to his face in anger. "That's none of your business," he told Maddox.

"You made it my business when you submitted their names," the General pointed out. "What is this?"

"They're as good of candidates as anyone," Jerry argued. "And they won't back out."

"How do you know that? You aren't asking them to go with you to Disneyland. This is another planet. Women are naturally protective, especially of their children. This has 'mess' written all over it…"

"Sir, I will take full responsibility," Jerry assured him.

"You better," Maddox said. "Does Madison know about this yet?"

"No," Jerry told him. "Not yet." In fact, Calvin's office was where Jerry planned to stop next.

"Jerry!" Calvin yelled when Jerry told him what he had just finished telling Maddox. "I know Candy. I don't see this working…"

"Trust me," Jerry pleaded. He had expected this kind of skeptical reaction from the General, but not from Calvin.

"And how did you get Lexi to agree to this?" Calvin continued. "Forgive me, but you and I know that Lexi is one of the most selfish—"

"Stop!" Jerry cut him off before he said something they'd both regret.

Calvin took a deep breath and approached the subject from another angle. "I'm just saying one or two days without cable TV and a mall to go to and the girl just might lose it."

"Stop! I will handle this. I just need approval."

Calvin threw his hands up in surrender. "Fine! Fine! But that still leaves us two short."

"I've asked Ben to join us. And I'm going to ask Major Blake."

"Blake?" Calvin asked. "You didn't even want her involved."

"I've had a change of heart. I believe she may be helpful."

"Jerry, didn't the ThreeofUs specifically say only one military personnel could be on the team?"

"I know. I'm going to talk to him."

Pinching the bridge of his nose and closing his eyes, Calvin reared his chair back in frustration. "Oh this is good," he said. "I see. So you don't just disobey orders and instructions on your own world, but it appears you plan to continue your tradition on into the next world and possibly the afterlife. I wonder what God is going to think about that."

Jerry didn't even crack a smile at the idea. "Calvin, I'm really glad that we had this conversation. I never knew you could be so humorous."

"Me neither," Calvin said before stamping his seal of approval on the candidates' files. "Let me know how this works out for you."

Wilkinson Residence — Blacklick, Ohio, U.S.A.
Earth Date: November 13th, 2014 — 7:59pm EST

"Kendra!" Barbara Wilkinson called up to her daughter as she set the dining room table. "Let's eat, Baby!"

There was a rumbling of footsteps as Kendra came barreling down the stairs and rounded the corner of their contemporary home, which was currently filled with the pungent aroma of delicious, soul food.

They sat down and, as usual, turned on the television to catch up on the day's news while they ate. "*Breaking news tonight,*" clamored the lead anchor of UNN as the eight o'clock news program began, "*as President Duffy is expected to address the nation momentarily. We now go live to that broadcast.*"

"Oh, what in the world is going on now?" Mrs. Wilkinson grumbled as she dished out food onto Kendra's plate. On the television, the shot of the news studio cut over to a live feed of the White House pressroom.

"*Ladies and Gentlemen, the President of the United States,*" a spokesperson for the administration announced from the podium, and Kendra turned up the volume.

The spokesperson stepped down and was soon replaced with President Duffy's tall frame and hard face. "*My fellow Americans. For years we have watched television shows and movies that show the possibility of visiting other planets and life on other worlds. I know that this will come as a shock to some, but tonight I am pleased to inform you of a unique opportunity that has presented itself.*

"*Approximately eight months ago, an astrophysicist at one of our leading establishments discovered and built a device that has opened a doorway of sorts to a parallel world very similar to our own. In many ways this world or planet is almost indistinguishable from our own planet.*

"*The atmospheric conditions and temperatures are very similar to ours, and although there are some unique differences, we believe that this world is an opportunity for our country to expand the peace and prosperity that we have come to enjoy and love as Americans.*

"We have spent the last six months researching the safety and sustainability of our gateways, and have concluded that though there are some risks, it is a risk that can be given as a choice to the American people. Just as many from foreign lands take risk to come to our country in pursuit of the American dream, once again a similar opportunity is available under the sanctions and auspices of the U.S. Government.

As I stated, this new opportunity is not without risk, so this is to be looked at as any investment opportunity.

"In one week from today we will begin accepting applications from everyday citizens to visit and possibly relocate to this new planet called Dabar. I realize, this can sound exciting and scary all at the same time, and trust me, this is not for everyone.

"Your local states will be providing you with information in your local newspapers and on the internet with the specific rules and stipulations to determine if this may be right for you. I believe this represents a grand opportunity for our people and our country. As I prepare to exit this office and President elect Slater enters, I am happy to have been able to play a small part in initiating this monumental piece of our history. Thank you, and God Bless America. My Press Secretary will take questions."

The President stepped away from the podium and Kendra could hear the reporters in the audience erupt with questions. "What in the world is this mess?" her mom said, confused. "Talking about going to other planets... This has got to be a hoax!"

"Mom, I think it's real..." Kendra said, stunned. Her mouth still full of food. "It's the President."

"Well if it is, I tell you what," Barbara Wilkinson proclaimed, as she began digging into her dinner. "I'm staying my black ass right here. We ain't got no business on no other planet. If the Lord wanted us on there, he would have put us over there already."

Kendra wasn't listening however. She was still staring at the TV, straining to hear the answers to the reporters questions. She couldn't believe what she was hearing.

Chapter 7

The Venetian Hotel & Casino — Las Vegas, Nevada, U.S.A.
November 14th, 2014 — 2:12pm PST

The ground fell away from Maychoff's feet swiftly, as the glass elevator soared upward, higher and higher, to the top of the hotel. Finally, there was a *ding* and the doors slid apart, opening to a lavish penthouse suite. Maychoff stepped off, escorted by two Secret Service agents.

Two more agents awaited them in the penthouse, and as the elevator doors closed behind them, one told him to raise his arms. Maychoff complied and let the man frisk him thoroughly. "He's clean," the agent announced.

"I know," a woman's voice said, as President-elect Patricia Slater entered from one of the suite's other rooms. "He's a friend."

She and Maychoff met in the middle of the room and shared a brief hug. "Bernard! How are you my friend?" she asked him, gesturing to join her on the sofa. "Please forgive the pat-down. It comes with the new job."

"No apologies necessary, Madame President-elect. Congratulations!"

"Thank you," Slater said with a modest smile. "You know, I must really like you for me to take a meeting on my vacation time. After I take office, I don't know when I'll be able to do this again without costing taxpayers an arm and a leg."

Maychoff returned her grin. "So conscientious. This is one of the many things I love about you."

"I have to be in this economy. I know firsthand what it's like to not have enough."

"I'm hoping I can help you with that piece of your job," Maychoff told her, "and thereby seal your legacy as the President who brought the U.S. economy back to life."

The two old friends sat quietly for a few seconds, staring at each other. Then Slater lifted a hand and made a motion that made her Secret Service detail disappear. When they were finally alone, she continued their conversation. "Really? If you can do that, perhaps you ought to be President."

181

He laughed at the idea. "No, no, no. I have other things to attend to, and I can be a greater help to you outside... Are you aware of the Dabar situation?"

"Just what's been shared by the President and the news coverage. It all sounds like a movie."

"Trust me," Maychoff said, all laughter gone from him now, "it's not. Now, I did not tell you this before you won, in the event that someone attempted to question our relationship. I wanted you to have no knowledge of this. The company that made the discovery, Troyconics Astrophysics, was mine until very recently. A few months ago, the United States Government bought me out. This new planet is very similar to ours and extremely rich in resources and is virtually undeveloped. It represents an opportunity to almost create another United States without any of the missteps."

Slater arched an eyebrow, intrigued. "What do you mean?"

"Patricia," Maychoff said, "I am talking about an opportunity for people to create and develop things that they have always dreamed of. And not just the wealthy and privileged. Anyone! And just like any investment, there are cost and risk associated, but the risks are low if rules are followed. It could be what you Americans would call a 21st century gold rush."

The President-elect leaned back in her chair, putting her feet up on a nearby ottoman. "Interesting..." she said. "Tell me more."

And so Maychoff laid out for her, everything that he had come to learn over the past months. About the Dimensional Mirror, about Dabar, about the Us. And finally, he explained his plan.

"Bernard, this is almost unbelievable," Slater asked when he was finished. "Are you sure this is safe?"

"I assure you," he told her, "this is all real and every precaution possible is being put in place to make it as safe as possible. Once the training program is in place, the risk will be greatly reduced. We can charge a modest fee for U.S. citizens—perhaps between $250 and $300. They are already duplicating the Dimensional Mirror per President Duffy's orders. Once word gets out, it won't be long before the Russians, the Chinese, the

Japanese and others will want in on the program."

"Do we really want the technology in their hands?"

"It's an entire planet," Maychoff pointed out. "You section it off and divide it, just as this planet is. The best part is I recommend a charge to the other countries of two billion dollars, and a percentage of the revenue generated by each country on the new planet. This will serve as a constant replenishing source of U.S. finances."

"Well Bernard," the President-elect said, sounding impressed. "It sounds like you have it all figured out. If this all pans out, I might as well put you in charge of overseeing all development on the new planet."

Maychoff flashed a winning smile. "I was hoping you would say that."

"We would just have to get you past the confirmation process. You still have dual citizenship, right?"

"Of course."

"Good," Slater said, taking a sip from a cup of coffee. "Keep your nose clean and this should be easy.

"Interesting that you should bring that up," Maychoff said, shifting in his seat. "It appears that 'my nose' is already dirty... Or at least someone believes it to be."

"What do you mean?" Slater demanded.

"It has come to my attention that for some reason I have attracted the attention of your Central Intelligence Agency."

"For what?"

"I do not know. But I assure you. I am clean. We discovered a mole in one of our companies in London, who has been feeding information back to the agency. Perhaps you can assist me in cleaning this up?"

"I'll find out what I can," she told him. "But Bernard... I need you to be clean with *me*. On everything."

"I have and I am," he assured her. "Perhaps this is all some type of misunderstanding. Everything we are doing has been above board."

"Good," Slater said. "I'll be in touch as soon as I get back into town. We have to be clean. The American people didn't put me in office for more of the same.

Maychoff stood up, buttoning his suit jacket. "Everything we have discussed will only make things better for your people and your country. Perhaps the world. You have my word." And with that, the two of them shook hands, hugged, and parted ways.

WORD WORLD

Jergensen Residence — Westerville, Ohio, U.S.A.
Earth Date: November 16th, 2014 — 8:23pm EST

Lexi lay on her bed, secluded in her room, and talking in hushed voices on her cell phone. Her father had made her promise not to talk about the other planet with Kendra, but what did he expect? She couldn't keep a secret this big without at least telling her closest girlfriend. And anyway, the President had already announced the news, so it couldn't do any harm now, could it?

"Soooo?" she asked her best friend. "Did you think about it?"

"*Yeah,*" Kendra said through the phone. "*I did.*"

"And?"

"*Girl,*" she sighed, "*look, you know I love you. You my girl. But I still live with my mom and she ain't havin' it.*"

"Why not?" Alexandria asked.

"*Lexi! You see the movies. When people are about to die, the black people know it and they don't go. Like Jason, Halloween, Freddy Kruger, all those movies. Girl, I can't do it. We ain't pose to be leaving the planet.*"

"Kendra! So you're just gonna let me go, but you're my girl?"

"*Girl, I can't stop you. Your parents want you there. But I'ma be praying for you though I already put yo name in the prayer box at church, so you gon' be alright.*" Lexi listened to the words coming through the phone, but said nothing. She was going to miss her friend. After a long silence, Kendra finally said, "*hello?*"

"I'm here."

"*Are you okay?*"

"Yeah, I'm fine," Lexi said. "Let me call you back."

"*Okay. We good, right?*"

"Yeah, we're good," Lexi told her before hanging up.

She stared up at the ceiling for a moment, collecting her thoughts. Then, she rolled off the bed and walked down the hallway to her parent's bedroom. The door was open, and her mother and father were both sitting up in bed, reading.

"So when do we leave?" she asked.

"In three days, Sweetie," her father told her. "Are you still in?"

Lexi gave herself one last chance to quickly think it over before making her final decision. "Yeah," she said. "I'm in."

"Are you okay, Lexi?" her mother asked her. Apparently everyone was suddenly worried about her.

"Yeah Mom," she said. "I'm fine."

Troyconics Astrophysics — Columbus, Ohio, U.S.A.
Earth Date: November 20th, 2014 — 9:28am EST

Three days later, Lexi found herself at her father's workplace. Though she had seen the facilities several times, there had always been places where she wasn't allowed. This morning, however, she was led to one of those places. Through a door with a *'Restricted Access'* sign, she was led through a winding maze of hallways and into a conference room.

Inside, the room was filled with strangers. With the exception of her parents, who had come in with her, Lexi didn't know anyone. Most of the people in the room were older than she was, but there were a few small children. Lexi studied their faces shyly as she took a seat at the end of the table.

"Mom!" a voice called from the hallway, "I'll be fine!"

Just then a young man, handsome and athletic, entered the room. His face was red with embarrassment and soon Lexi saw why. His mother followed him into the conference room. "Okay, Baby. Did you pack your toothbrush?"

"Mom, I have it," the boy said.

"What about your night socks? You know how your feet get cold—"

"Mom, you're embarrassing me. I love you. I'll be back. Goodbye." He kissed his mother on her forehead before gently pushing her back into the hallway and closing the door to the conference room. Once she was gone, he turned back around, spotted an empty seat next to Lexi and sat down beside her.

"Hi," he introduced himself. "I'm Josh. Sorry about my mom, she's a little over-protective. Glad to see there's at least one person here that's my age. And you're cute too."

Lexi felt her face go red. "Yeah," she managed to stammer. "Me too."

"Okay, if I can have your attention please," a large man in a perfectly tailored military uniform said from the head of the table. "My name is General Maddox, and I am in command of this facility. Now this will be a briefing with instructions for your trip, so listen up closely. You've all been health screened and given some preliminary information as well. We will be providing you with more detail during the briefing but first we need to do roll call…"

Dimensional Mirror Insertion Point — Dabar
Date & Time Unknown

The briefing had gone by quickly as did the prep time to get ready to go thru the portal to Dabar. Since they assessed no suits or helmets were needed this sped the process along as well. What had not gone over well was the until now relayed fact that none of the civilian team was allowed to speak once they were on the new planet. Jerry and Calvin spent nearly an hour reassuring the assembled team that once the needed precautions were in place this stipulation could be lifted. Although still skeptical, the team allowed themselves to be persuaded with the help of some monetary incentives. The monetary incentives were General Maddox idea, as he had no desire to see the trip delayed any longer. After last minute checks and the firing up of the Dimension Mirror, Jerry led the wonder-struck team through the Dimensional Mirror and into the lush countryside of Dabar. It somehow seemed even greener than before. And there was another change. As they all cleared the portal, Jerry saw four brand new Hummers parked in the grass just twenty yards from them. He walked up to them, peeking through the windows and spotting the keys lying on the driver's seat. He smiled. The Us had left these for them as gifts.

"Alright everyone," he called back to the group, who were all staring at their new world with wide eyes. "Load up a vehicle and follow me. Major Blake, can you drive one vehicle?"

"Sure," Blake said.

"Candy, if you can grab the other one. And you, Tamyra."

"Alright," Tamyra said with a smile. "This is turning out to be a pretty sweet deal already. I got my own Hummer!"

Together, they packed all of their gear and supplies into the roomy trunks of the SUV's, then climbed in and drove off, headed for the Sanctuary. The SUVs bounced and rocked as they climbed over the terrain, but as they got close to the castle, suddenly the ground leveled off. It was paved.

That's when Jerry saw the alterations that the Us had made. The

Sanctuary was not the only structure on Dabar anymore. Just next to it, there was now a cluster of buildings that Jerry could only think of as a hotel or resort. He could see a pool, a playground, a garden, and both tennis and basketball courts. They stopped the Hummers in front of the resort and got out to find the ThreeofUs waiting to greet them.

"Good morning, Jerry," he said with a slight hint of cheer.

"How did these trucks get here?" Jerry asked.

"We created them the day you left, along with most of this resort. Hopefully it is to everyone's liking?"

The team stared at the ThreeofUs, then glanced over at the resort, nodding their heads in stunned silence.

The ThreeofUs maintained his usual emotionless face. "Human beings have a tendency to long for normalcy. We have just finished the rooms however, and there is a bit more that needs to be done. We were not aware of all your personal preferences until you came through the mirror portal."

Candy leaned over to her husband, not taking her eyes off ThreeofUs. "He looks like us... but he's a kid" she whispered in Jerry's ear.

"Looks are deceiving," he whispered back. "Now shhh! No talking. You don't get special treatment just because you're my wife."

"There is no need for concern right now, Jerry, as to you and your companions speech, as we have extended the parameters of the Grace Asylum. This area, as well as your living quarters are now under Grace," ThreeofUs informed them.

Jerry smiled, surprised and relieved all at the same time. "I didn't realize you could do that."

"There are so many things we all do not know. Even Us. If one is able to constantly acknowledge this throughout their entire life, that person will never cease to grow."

"I'll be sure to keep that in mind," Jerry said, before turning back to face the group. "Everyone! This is the ThreeofUs." He then took a few steps forward, whispering to the ThreeofUs under his breath. "Hey listen, maybe don't do the whole awareness thing, and call everybody's names before I say them. I don't want to freak people out. Some of them are still a little edgy."

"We were aware of your desire, Jerry, as well as their apprehension. We had no intention of doing so."

"Oh, I knew that," Jerry said feeling a little silly. "I was aware…" Then he stepped to the side and began introducing the team to the ThreeofUs. "ThreeofUs, this is my wife Candy, and my daughter Alexandria is over there. This is Major Keira Blake."

Blake stepped forward and extended her hand bravely, hoping to shake hands with the alien being. The ThreeofUs simply stared at her then at Jerry with the slightest hint of agitation, however, refusing to shake, and soon the Major took her hand back, confused.

"Can we talk about this later?" Jerry muttered to ThreeofUs. "Alone?"

"We will," ThreeofUs proclaimed before letting Jerry continue with the introductions.

"This is Josh Maybran, and you know Field Specialist Tamyra Ann. And over here we have Samantha Morgan and her two twin boys, Max and Mack."

ThreeofUs bent over at the waist, bringing himself down to eye level with the two little kids. "Young ones," he said. "Good."

"And then there is our summer intern, Sanja Khemdesh, and one of our employees, Lisa Hansley. And last but not least, this is Ms. Jankowski."

An older woman stepped forward from the back of the group, the wrinkled skin of her face stretching into a grin. "They say you can't teach an old dog new tricks," she said. "But I told Jerry you can if you have a good teacher."

"A wise woman," ThreeofUs retorted. "And still full of much life. She will prove to be an asset and a great example. You are how young?"

"Seventy-four," Ms. Jankowski said.

"Good," ThreeofUs told her. "Just slightly beyond middle age. And with that, he turned around and began walking away.

A little confused, Jerry ran after him. "I wanted to ex—"

But ThreeofUs turned quickly to him, and staring deep into Jerry's eyes, reprimanded him. "You are a good man, Jerry Jergensen. But you have learned enough and seen enough to know that insubordination here on

Dabar is costly. It is just as costly on Earth, but due to the fact that the consequences are delayed deliberately, many continue in their ignorance. Your presumption that things on Dabar can be handled in this manner is dangerous. The number twelve is a specific number with a specific reason and a specific meaning. There is a reason I told you twelve people. But We will allow the thirteen… At this time. Have We made Our future expectations of your behavior clear?"

Jerry sighed, defeated. "Crystal."

ThreeofUs nodded. "Please do not disappoint." Then, as if seeing her for the first time, ThreeofUs looked over at Lexi and walked to her. "Hello. You are Jerry's daughter, Alexandria."

Lexi was speechless. She simply stared at the being that looked as if he was the same age as her. "It's okay, Lexi," Jerry told her. "You can talk."

She looked over to her Dad and took a deep breath and released the barrage of all she had been thinking and withholding with literally no breath between sentences. "Okay, Dad. Which one is it? First you and the General say we can't talk, then you say we can. And Dad, that was so not right for you to not tell me about the whole no talking thing until we were actually getting ready to go. You know I would have never agreed to come here if I knew we couldn't talk. I mean, talking is like, my *thing*. That's me. I talk."

The ThreeofUs continued his stare at her; his eyebrows rose, and then slowly turned to Jerry. "Perhaps it is best if this one does not talk."

"Very funny," Jerry said.

"Trust us," said ThreeofUs, "Humor was not our intent…"

Jerry turned back to the rest of the group. "Okay everyone! I realize we pounded the idea of 'do not talk' into you before we left. The reason we did is because talking is so normal that it's hard not to do. We were not aware that an area would be provided where it would be safe to talk. So everyone stay in this area only and it is safe to speak in this area. Do not wander outside of this area for any reason and if you do, *do not speak*. After we unload, we will let you know the perimeter which you should not cross."

Everyone nodded that they understood. Then Candy broke off from the group and went over to the beautiful garden next to the resort. She began examining a rose bush there, and soon was joined by Major Blake and Ms. Jankowski.

"Oh my!" Candy exclaimed. "These are the most beautiful roses I have ever seen!" She reached out and felt one of the vibrantly red petals as Jerry and ThreeofUs joined them at the garden. "Are these real? They don't have any thorns on them."

The rest of the team began walking over too, crowding around them. "Understand," ThreeofUs answered. "Everything here, unless it becomes perverted, is manifested in its true image and original design."

"What do you mean?" asked Candy.

"The roses were always truth," the ThreeofUs exclaimed. "The thorns were always a lie."

The humans looked at each other, confused. Only Jerry, Candy and Major Blake seemed to grasp what the ThreeofUs was saying but even then, only vaguely. "There is much to learn," ThreeofUs continued, "but for now We will show you to your living areas."

As the ThreeofUs began to lead the way, Jerry fell into stride with him. "Were the vehicles to your liking, Jerry?" he asked.

"Yes," Jerry replied. "I have always wanted a Hummer."

"We know. My apologies that it was only a 2008 model. We had to work with what We knew and We have not been to Earth since 2008. It was the last model We have seen."

They entered the resort, and ThreeofUs took them down a long corridor that resembled the hallway of a five star hotel. "Everyone's room has been designed to their preferences and likes," ThreeofUs explained. "In some cases we have tried to duplicate—"

Jerry nudged the ThreeofUs gently. He was worried the being might be getting a bit too mystical for the group's taste again. It seemed to work, because the ThreeofUs stopped speaking, and rephrased what he was about to say: "We hope they are to your liking." Then ThreeofUs spoke in Jerry's ear. "Was your assault on Us an attempt to get Us to lie?"

"No," Jerry said. "Just to get Us to adjust what Us was about to ay."

ThreeofUs nodded. "We are relieved. Because if there is one thing you must understand about Us, Jerry, it is that *we do not lie!* It would destroy Us."

Jerry looked back at the ThreeofUs. "I understand."

"Good! Now you should advise your companions to settle in and get some rest. We will all meet at the archway entrance in the morning and begin our training. It will not be what you expect but it will be effective. I will see you at that time." And with that, ThreeofUs headed back towards the Sanctuary.

"Everyone, lets get settled into our areas and get some rest. You have your food packs and other materials. We will all meet out in front of the castle so everyone synchronize your watches. Be out front no later than 0900. Nine A.M. for all of you non- military personnel.

Then, Josh Maybran asked a strange but sincere question. "When does it get dark here? And where is the Sun?"

The Sanctuary of Us — Dabar
Date & Time Unknown

The next day, the group followed Jerry's instructions and assembled at the front entrance of the castle. As soon as everyone had arrived, the ThreeofUs once again stepped out of the blackness of the rear archway.

"Good," he said. "Promptness and being on time are part of the training. It is good to see that everyone has passed this first test. Follow Us."

They followed the ThreeofUs inside through the interior of the front room of the sanctuary thru the far archway that Jerry had not been able to pass thru. The group murmured with wonder as they saw the furniture on the ceiling and the staircases that wound sideways along the walls. The ThreeofUs took them past the terrace where he and Jerry had first talked.

Pausing his measured gait the ThreeofUs addressed the proverbial elephant left in the room by Josh Maybran. "As many of you discovered there is no night time here. In an effort to compensate for your unnatural need for this absence of luminance your quarters were all designed with room darkening shades. In addition, there is also no Sun here," he said as if such things were as normal as breathing.

Josh Maybran piped up, "So where is the light coming from? And the heat or warmness?"

"The explanation of that inquiry at this point would prove counter productive to our training," the ThreeofUs relayed. If you would allow Us to forgo answering it We would be most appreciative. As a matter of truth, if you focus and listen to what is taught you will be able to answer that question yourselves." Everyone nodded their approval but their eyes still longed for the answer. They would have to wait.

Finally, he led the group to an unimpressive large room with a chalkboard and rows of school desks. There were exactly thirteen desks, and each one had a notebook, pens, pencils and what looked like a large green textbook on top of them.

"As Jerry was told previously," ThreeofUs announced. "the main thing that you must learn is to unlearn what you have learned. This is why you are so surprised at very simple things such as staircases that are not just up and down but also sideways as well as furniture that is stationed on what you believe is the ceiling and not the floor. For those of you with a faith background, your faith will assist you in mastering the skills necessary to live a successful life here more quickly. For those of you who believe otherwise, your learning curve here will be…different."

"Where are the computers?" Alexandria asked.

"I am glad you asked that, Alexandria," ThreeofUs said. "Unlike what has become of schooling on Earth, all you will need to train here you already have. Primarily your ears, your eyes, your mind, your imagination and your mouth. The note pads and pens are there for you to simply record and review what you have been taught. Although you have it recorded in your memory, you have not yet allowed yourself to discover how to access all that you have ever learned, heard, seen or ever known."

"Is that possible?" asked Candy.

"Again. We would not have said it if it were not. As a people who are accustomed to lying and being lied to, it will be good for each of you to dispense with that fatal habit. Things have a way of popping up so to speak here, *even if they only exist as a lie.* Here you are free to ask any question and not just receive an answer, but to receive the truth. Do not forget we have told you this."

Josh Maybran lifted the green book on the desk closest to him and felt its weight. "What's up with the fifty pound dictionary?" he asked.

"I'm glad you asked," ThreeofUs said expressionless. "On each of your desks is a Webster's 1828 dictionary. The reason for the vintage of the dictionary is due to the truth that certain people and influences, who will remain unnamed, have for their own purposes and advantages attempted to adjust the meaning of words to fit their own…agenda.

Little Max Morgan raised his hand and ThreeofUs looked at him kindly. "Yes, a properly presented question. Yes Max?"

"What does vintage mean?"

ThreeofUs turned to look at the other twelve persons. "Does anyone care to answer the young man?"

"It's the year a wine was created or produced," Jerry answered.

"Jerry is correct," ThreeofUs said, "in part. On your planet the word 'vintage' is used in majority in regards to wine. However, it also means the time that something of quality—high quality—was produced." He took one of the dictionaries from another desk and held it up. "This is high quality. Without it, one can lie without knowing it, or speak in error. And here on Dabar, as on your world, regardless of ignorance or error, both can be dangerous. Please, be seated."

The group milled about, each person choosing a desk and sitting behind it. When they were all settled in a chair, ThreeofUs made his way to the front of the classroom and officially began their first lesson. "So as this…changing the meaning of words as time passes continues, understand that in truth this can not be done and even though it is done, the change only exists as a lie."

"What did he just say?" Lisa Hansley muttered in the back of the class.

"I don't know either," said Sergeant Barclay, confused.

"For those of you who, like Sergeant Barclay and Ms. Hansley, have no clue as to the truth that we just spoke, fear not. We will elaborate upon it later. Now, tell Us: what is a word?"

"A way of communication," offered Major Blake. "By combining different ones, we communicate."

"Yeah," Lisa said. "A way to talk or communicate."

"Communicate what?" ThreeofUs prompted them. "More words."

"No," Blake said. "Well…sometimes."

The ThreeofUs walked forward until he stood over the seated Major and stared into her. "Dog," he said.

"Excuse me?" said Blake.

"We said 'dog.' When We said 'dog,' did you see a word, or did you see a picture?"

"A picture," the Major answered.

"Of what?"

"My dog."

"Big dog," the ThreeofUs went on. "Big black dog. Big black scary dog. Big black scary dog running after you." When we said these words, did anyone see words in their mind?"

The team all shook their heads, murmuring no's as they began to understand what the ThreeofUs was getting at.

"What did everyone see or imagine?"

"Pictures," Lexi said.

Candy nodded. "A big black scary dog running after me."

"Does everyone agree?" ThreeofUs asked the class and they all nodded in unison. "Good. Class dismissed. We will see you tomorrow at the same time. The same place."

"Is there any homework?" asked Jerry.

"We have just given it to you," ThreeofUs responded. "Practice it."

"Practice what?"

"Check your notebook, Jerry."

Jerry flipped open the cover of his notebook and stared at the first page. At the top, in perfect penmanship, the page read: *Practice understanding that words create pictures.* Ooohs and ahhhs filled the room as everyone else checked their notebook to discover the assignment already written there.

Date & Time Unknown

The next day, everyone was once again on time for their lesson. Everyone with the exception of Sergeant Barclay that is. The Sergeant ran into the classroom five minutes late, looking out of breath and embarrassed. The ThreeofUs stared at him as he found his seat.

"We see you, Sergeant," he said with a warning tone before turning to the rest of the class. "Did everyone complete their homework assignment?"

The class nodded and mumbled their yes's.

"Good. Because in order to progress further you must *believe*. If you do not, then We are going to ask you to leave right now." As the ThreeofUs said this, he continued to stare at Sergeant Barclay, as if the invitation to dismissal was really meant for him and him alone. Everyone however remained seated. "Very well," the ThreeofUs finally said. "Tamyra?"

"Yes sir… I mean yes?"

"Did you remember to open the envelope We gave you on your last trip here?" ThreeofUs asked, as Jerry suddenly remembered that he had forgotten all about the envelope. He had never opened his.

"Yes, I did," Tamyra said.

"Good."

Not wanting to be embarrassed in front of the rest of the class, Jerry quickly groped inside the pocket of his field pants, and pulled out the envelope. He tore it open to find a folded piece of paper inside. However, when he opened the note, he saw that it was nothing but a blank piece of paper.

"Do you have the item We requested?" the ThreeofUs said, still talking to Tamyra.

"Yes," she said, reaching into her backpack and pulling out a boxed LEGO set. She handed it to ThreeofUs and Jerry could see that it was an R2-D2 set.

"We are positive that everyone knows what this is," ThreeofUs explained. "But, how many of you know what the word 'LEGO' actually means?"

"It's a variation of the word 'logos,' which means 'the word.'" Major Blake offered.

The Major is correct… in part," ThreeofUs said. "The word LEGO in its sense is a Danish word from the 1950's, which means to play well. However, the original meaning in Hebrew is 'I say,' or 'I speak.' It is no coincidence that the name used for this toy represents and is associated with words. Just like LEGO are used to build something—in this case R2-D2—words are used to build things. In Truth, words are used to build everything. Everything you see on Dabar and on Earth was constructed, created or manifested by words. Although our initial focus will be to teach you about words in English, there is even greater power and ability found in using the Hebrew language here. And if one masters that, nothing will be restrained from you that you have imagined to do. If you do not understand this or believe this, there is absolutely nothing you can do on this planet except lie, because everything here that is real or truth is controlled by words.

"Class is dismissed. Your homework is to decide what you believe and then decide whether you need to stay. I will see everyone on time tomorrow. Anyone later than one minute will not be able to enter the room."

And with that, ThreeofUs gave a pointed stare at Sergeant Barclay then turned to leave the classroom. Jerry, however, still had questions, and he jumped out of his seat to intercept ThreeofUs at the door. "Yes Jerry?" ThreeofUs asked with his back still turned to Jerry.

"I just wanted to apologize for forgetting to read the letter," Jerry told him. "But it was blank anyway, so I guess it didn't really matter."

"Please know that We have begun to expect more from you, Jerry," ThreeofUs told him. "What you must realize, understand and ask yourself is this: did We give you the blank paper because We neglected to write something on it? Or did We give you the blank paper because We knew you would forget about it?"

As the ThreeofUs walked away without another word, Jerry stopped, thinking about ThreeofUs words as he continued down the corridor and out of sight. He stared at the blank paper and came to a conclusion. "They knew I would forget about it." He said to himself, balling up the paper in hand. As he exited the corridor, still deep in thought, he noticed a wastebasket on the ceiling of the front room he just entered. Without thinking he tossed the paper wad toward the wastebasket. Defying Jerry's knowledge of gravity the wad shot upwards and into the basket and stayed as if it was right side up. Jerry was baffled and scared. He increased his pace out of the room turning back once to stare at the basket.

Troyconics Astrophysics — Columbus, Ohio, U.S.A.
Earth Date: November 24, 2014 — 2:14pm EST

Back on Earth, Calvin found himself sitting in General Maddox's office. He had been summoned for a meeting (though he didn't know what it was about) and now, as he listened to the General talk, he began to realize how much he wished Jerry was not still on planet. He was far better at dealing with Maddox than Calvin was.

"The additional gates have been manufactured," Maddox told him, "and they're currently being set up for tests. Although our people have been working along with Troyconics employees for a few weeks now, they still have not mastered the physics and operation of all the equipment. So…"

"So what?" Calvin asked, apprehensive.

"So we're going to need to fly a few of your people out to Nevada for a couple of days to test the new mirrors and make some adjustments."

"General, we gave you everything you need to duplicate what Jerry has created here. In addition, I don't believe it's wise to remove key personnel for the mirror's operation from this site, when we have people off world as we speak."

General Maddox frowned. "Madison, I understand your concern. But Jergensen and his team are not scheduled for return for three days."

"But what if there is an emergency?" said Calvin.

"We will leave the joint team members here," the General explained. "They are well versed in the operation of the Dimensional Mirror."

If they can operate this mirror then they should be able to operate the new mirrors as well… Unless of course you've made modifications."

The General's expression betrayed the slightest hint of a smile. "We have made some."

"Like?" Calvin pressed.

"Some of the mirrors are larger in size and dimension than the one currently here."

"Really? And why would that be? Are you planning to drive a semi-trailer through one?"

"Perhaps," said the General. "But that's up to the President. Listen Madison, I need Ian, Rajin and Sara Choremo ready to fly out at 0800 tomorrow to the base in Nevada. I also need you to have Benji Sabine on call as he may possibly be needed as well. I'll know by 0200 tonight. Thanks for your assistance."

Then Maddox raised his arm toward the door, inviting Calvin to leave. Calvin thought about arguing the matter, but the look on the General's face told him that would have been a bad idea. Instead, he left the office and went out into the hallway. There he leaned against the wall and took a deep breath. "What are they up to?" he asked himself aloud.

The Sanctuary of Us — Dabar
Date & Time Unknown

Bright and early, the new students of Dabar quickly made their way from the resort, to the castle next door. Each of them now knew their way through the sanctuary and into the classroom, where they each took their seats quietly and waited for class to begin.

"Good morning," ThreeofUs greeted them from the blackboard. "We will be outside today."

He took them out of the class, through more bizarre and impossible hallways, and finally brought them to a room none of them had seen before. It was a large atrium with a glass top, and the sun shone down through it, lighting up the entire room with cheery morning luminescence.

"We are sure," ThreeofUs, told them, "due to your initial impatient nature, that you have grown tired of the repeated and daily sessions of defining words. Trust Us, it is necessary. Today, we will actually begin to make use of what you have been practicing the last few days."

The celestial being with the teenager's body gestured to a corner, where one of the six-second boxes was mounted, dark and dormant. Jerry saw that there was a box in each corner of the room and there were two in the center. "Jerry is familiar with the silver boxes which line the corners and center of this room. Albeit he does not know exactly what they are."

"They're some type of warning device," Jerry said.

ThreeofUs nodded slowly. "That they are, but they are a warning device and a grace device. The 'six second boxes,' as Jerry has so affectionately named them, activate when words that will result in death and or destruction are spoken. A person has six seconds to rescind or refute that word spoken whether it was by them or toward them. Although Jerry refused to tell everyone here of the incident, three of his earlier team members died while speaking negative words, or as a result of them being spoken."

As he remembered the unpleasant incident, Jerry felt the flames of anger rekindled in his gut, getting hotter and hotter. "And the ThreeofUs neglected to mention that he—or They—watched the whole thing happen and did nothing."

"Jerry," ThreeofUs said in his usual neutral tone, "your attempt to diminish Us in the sight of your team is not admirable. We meant no ill will or spite. You on the other hand have allowed an emotion to take control of you and thereby your words. This is a mistake that can cost many their lives.

"Our only reason for informing your team members of the incident is to give them a very conscious understanding that death is a choice and an option. Specifically if they do not learn to control their emotions and better still to remove the emotions from words and focus only on *truth*!

"If Jerry were to focus on the truth, he would have to admit that each person, whether intentional or not, killed themselves and others. But in order to accept this truth would require a character trait that the people of your planet are increasingly lacking. *Personal responsibility*!

"Unfortunately, six thousand years later this character flaw which first surfaced in your predecessor Adam, if you believe the Bible is alive, well and still persist today. The tendency to blame others for one's own choices, shortcomings and failures must be eradicated to be completely successful here. Excuses are for the weak!"

The ThreeofUs looked around at the class, who were all staring back at him in stunned silence. He smiled. "Now that We have your attention, let's begin today's hands-on training. Sergeant Barclay?"

"Sir!" Barclay barked instinctively before realizing who he was addressing. "I mean—yes?"

"What was the name you called Us in your head upon your first visit, when We declined your entrance into the sanctuary?"

Barclay looked around, unsure of whether or not he should answer. "I called you, uh… I called Us an asshole."

Beep! The six-second box closest to Barclay activated and lit up with the number 6. As the countdown began, the class gasped in alarm. Barclay's eyes went wide with fear and he reached for something tucked in the waistband of his pants, but—

"Fear not!" the ThreeofUs cried out. "You are all under Grace!"

They waited nervously as the box finished its countdown. It hit zero and then went black once more. Sergeant Barclay remained untouched.

"One of the greatest deceptions most of the people on your planet live under," ThreeofUs explained, "is that their words mean and do nothing.

Nothing could be further from the truth! The truth is here on Dabar, if We were only human and Sergeant Barclay called Us the name he stated and We did not refute it or refuse to receive it, We would immediately begin to become an irritating and contemptible person to ourselves and to others."

"Understand, this would have happened not because We are irritating and contemptible persons, but only because We believed and accepted what the person called Us and therefore We became that even though it exists only as a lie. Trust Us. This has happened millions, billions of times to many on your planet.

"As you sit here, you know of many people who were called failures and disappointments and nobodies by parents, classmates or co-workers. They unknowingly accepted that call and today, even if they are not failures in the world's eyes, they possess a failure mentality."

The class finally seemed to grasp the concept the ThreeofUs was trying to relate to them. They bobbed their head in understanding nods as the being went on.

"Even on your planet when a person is called by the name Sergeant Barclay so casually called Us—although mentally—specifically on a repeated basis if one were to pay attention they would notice how much more irritating and contemptible that person actually becomes to them. Questions?"

Candy raised her hand. "How is it that these changes do not take place immediately on our planet?"

"Because the power of your words has been…turned down," ThreeofUs explained. "Way down. And the harvest of the word seed you speak has been severely…retarded."

"By whom?" Candy wanted to know.

The ThreeofUs walked over to her, staring deep into her eyes. "Dear Candy. Why do *you* ask questions about that which *you* already know the answer?"

Candy looked down at her feet then back at ThreeofUs and whispered, "I just wanted to be sure."

The ThreeofUs turned back to the gathered class. "As We do not wish to insult anyone's religious belief, We will leave the whom to each person's own imagination and belief. Your homework is to figure out if you believe what We have taught you today. Remember, people have died for you to be here today. Do not take your being here and the opportunity you have been given lightly. Class is dismissed. We will meet you all here at the same time tomorrow. And Sergeant Barclay?"

"Yes?" he answered.

"Be sure to leave your toy at home tomorrow."

The Sergeant's jaw dropped and he subconsciously patted the back of his waistband, feeling the weight of the pistol he had tucked under his shirt. It was what he had been reaching for when the six-second box had started its countdown, but he hadn't pulled it out. And yet somehow the ThreeofUs had still known about it.

Date & Time Unknown

Moments later, as the humans left the sanctuary and made their way back to their resort, ThreeofUs watched over them. He stood unseen at the railing of a terrace, watching his students below. OneofUs stood next to him, looking at the same creatures and scowling.

"Is it that distasteful?" ThreeofUs asked.

"Indeed," OneofUs replied. "I have not sensed such disunity, such disharmony in a group of people in a very long time. It...is...painful."

ThreeofUs nodded as they stared down at the oblivious human beings. Then the two beings spoke together, perfectly in sync. "It is unfortunate," they said as one, "that there is more to come."

Date & Time Unknown

Lisa Hansley's room had been painted a deep lavender color—her favorite color—and it had a calming, soothing effect on her. Now, before it was time for lunch, she sat at the desk that had been provided and compiled her report for General Maddox.

"Today was thankfully somewhat different," she said into the voice recorder she'd been given. "We were informed of the reasons why certain words cannot be spoken and we were also told what the small metal boxes that are scattered throughout the planet are. Six-second boxes, he called them. Apparently when a person speaks words that could lead to death or destruction they have six seconds to refute or rescind them. If not, they quite possibly could die, as we were informed that some others have."

She clicked the button to stop the recording, then thought about it and added one last thing: "That would have been nice to know before we came here, by the way. That's it."

Then she powered off the recorder and climbed into bed for a nap.

Date & Time Unknown

The resort's dining room was a large, open-air space with a view of the countryside and a long wooden table around which the visitors to Dabar could sit. At the moment, the table was occupied with only four of the students. There was Samantha Morgan, her two sons Max and Mack, and Ms. Jankowski. Together they talked as they ate the packets of food rations the army had supplied them with.

"You know," Ms. Jankowski said. "I realized that when you have something all the time, it's easy to take it for granted."

"What do you mean, Ms. J.?" Samantha asked.

"Food!" the old woman said with a laugh, eyeing a forkful of the flavorless stuff. "I never thought to ask about the food. I just assumed we'd have the same food as we have on Earth."

"Oh!" Samantha said. "I know what you mean."

"I had often thought about joining the army," Ms. Jankowski went on. "But I see now why I didn't. This stuff they called food is horrible. I think I've lost fifteen pounds."

Samantha laughed. "They're called MRE's. My brother's in Iraq. He talked about them all the time when he first joined. He thought they were horrible too."

"My word. MRE must stand for Mess Rarely Edible."

This time they both laughed. Max and Mack however, didn't seem to find it funny. "Well the boys love them," Samantha said. "They think they are the coolest things."

"They just don't know any better," the old woman said with a wink.

Date & Time Unknown

The next day, they found themselves back in the classroom, notebooks out and ready to learn. The humans sat in their desks as usual and the ThreeofUs stood at the blackboard, dispensing knowledge with his signature mysticism.

"We are sure that many of you have not understood the monotony of us defining words," he began. "Some of you expected much more. Upon your returning, We will begin to actually move into the manifesting of things. Notice We did not say creating of things."

Then ThreeofUs turned to Ms. Jankowski and gave her a slight smile. "We would also like to apologize for our neglect of more desirable food choices. As we do not consume food, although we can, we allowed it to go beneath our notice, a situation that will be rectified immediately. However, take comfort that we have decided to allow you to return to your planet two days early. It has come to our attention that Thanksgiving is two days away and the ability to engage in this feast will more than likely restore some of the joy some may have lost during their undesirable diet here."

"Do you not celebrate Thanksgiving here?" Candy asked.

"Certainly!" ThreeofUs proclaimed. "One cannot truly live without being thankful. They can exist but they cannot live. However, We could never limit such a component of true life to just one day. Therefore We are thankful every day! Does anyone have any questions?"

"Yes!" called out Ms. Jankowski. "Can we get a Burger King? Maybe a Taco Bell and a KFC too? Hell, I'd settle for—

Beep! A six second box in the corner of the classroom suddenly came to life, counting down from six as always.

"I take it back!" Ms. Jankowski cried. "Heaven! I meant Heaven!"

The box reached its zero count and went dark. Ms. Jankowski looked around, almost as if surprised to still be alive. "Thankfully we are under Grace," ThreeofUs said. "However, do not train yourself to replace inappropriate language with a lie, because it will become a habit. You will be caught eventually and when you do, more inappropriate language is sure to result." Then he beamed down at Ms. Jankowski. "We will work on your request, Ms. Jankowski. Mrs. Morgan?"

"Yes?" Samantha asked.

"Would it be possible for you and your sons to stay briefly after the dismissal?"

"Sure."

"Good. Class dismissed. Hopefully I will see everyone who belongs here next week." The ThreeofUs seemed to be staring at Sergeant Barclay as he finished speaking, and Jerry wondered if that was significant.

Before he could read too much into it, however, the class was dispersing. The men and women got out of their seats and made their way out of the classroom, leaving only Samantha and her sons behind. As they filtered out into the hallway, Alexandria sidled up next to Josh Maybran.

"Thank God we get to go back early," she said to him.

"Yeah... I'm a little bummed though. I actually like it here."

"You do?"

He shrugged. "What's not to like? Learning new stuff. Nobody to mess with you. There's no crime here. It's like a whole new world. I love it!"

"I never thought of it that way," Lexi said.

"Oh well! My mom will be happy."

Meanwhile, back in the classroom, the ThreeofUs walked over to where Samantha and her boys still sat in their desks. He stooped down so that he was eye level with them. "We were interested in finding out if you would like to get a jump on your lesson next week. You and the young ones."

"Sure," Samantha said. "What do I need to do?"

"We were actually going to have Max and Mack join us as well. Hello Max. Are you having a good time?"

"Yes," Max said emphatically. "I love words. I love reading the dictionary."

"Yes," ThreeofUs nodded. "We know. We can tell you are having a good time. Tell Us, does your mother take you both to McDonald's?"

Both boys shook their heads 'yes'.

"Good. Can you close your eyes and do Us a favor?"

"Yes," Max said, and they both obliged.

"Imagine the McDonald's now in your mind," ThreeofUs instructed. "Can you do that?"

"I am right now," said Max.

Then the ThreeofUs reached out and pressed the thumb of each hand against the center of each boy's forehead, closing his own eyes as he did so. Concentrating, the ThreeofUs could see what the boys were seeing. He could see into their minds.

"Good," he said. "Now can you turn around in the McDonald's in your mind, so that we can see everything in it."

"Yes," Max said, eyes still closed. Inside their shared vision, he turned and revealed a 360-degree view of the building's interior.

"Thank you, Max," ThreeofUs said, opening his eyes and removing his hands from the boys' foreheads. "Thank you, Mack. You've been most helpful."

"I guess you all didn't need me…" Samantha said.

ThreeofUs turned to her. "Indeed we do. When you return, can you please visit Taco Bell, Bob Evans and other restaurants you frequent? Simply go in, look around and see everything you can see. This will help Us when you return."

"Sure," Samantha said. "No problem. You want the inside and the outside?"

"Certainly. We were going to create the outside from memory, but a more updated look will probably make everyone feel more at home."

"Works for me," she said as she got up. The ThreeofUs stopped her, however, noticing the crucifix that she wore around her neck.

"A cross," ThreeofUs said. "Tell Us, are you a cross wearer or a Christian?"

"I take my faith very seriously," she told him. "I *am* a Christian. Still growing, but a Christian no less."

ThreeofUs nodded. "Thank you Mrs. Morgan. And thank you Max. Mack, we will utilize your unique skill set when you return. Happy Thanksgiving."

Date & Time Unknown

Later that day, the team gathered outside the resort. The Hummers were parked by the entrance and all of their gear and clothes had been loaded inside. People milled about, eager to go home but waiting for the official order to get into the SUVs.

As Jerry loaded up the last of the supplies that couldn't stay behind, the ThreeofUs left the castle and approached him, another envelope held in his outstretched hand.

"Another blank note?" Jerry asked.

"No," ThreeofUs said. "Actually we require some things from you upon your return. If you appropriate these items for Us, We would be most grateful." The ThreeofUs then holds a closed hand to Jerry and as Jerry opens his hand the ThreeofUs drops a single, flawless five a carat diamond into Jerry's hand. Jerry hands it back to him.

"Can I open it now?"

"Certainly."

Tearing open the envelope, Jerry unfolded the piece of paper inside and read it. "You want pictures? Pictures of all these things and places?"

"Pictures in the least. Diagrams and blueprints would be better. Of your lakes, your sewage system and, of course, a few houses."

"May I ask why?"

"Having people here necessitates the need for many things We do not need," he explained. "We accommodate. It is never our purpose to cause undue tension or heartache. Especially over something as trivial yet vital to you as food. Happy Thanksgiving, Jerry. We will see you soon."

Chapter 8

Two nights before Thanksgiving, the White House was abuzz with hectic last-minute business and meetings. Deep in the bowels of the West Wing, President Drake Duffy had summoned the Joint Chiefs of Staff, along with President-elect Slater, General Maddox and top members of the intelligence agency, for a meeting of their own.

"Ladies and gentlemen," the President greeted them from the head of the table. "Thank you for taking the time to come out. I realize many of you are preparing to possibly head out of town or to your homes to spend Thanksgiving with your families. There are two reasons I've called everyone here. The first is to discuss our progress on the training and preparations of the new Dimensional Mirrors for Dabar. Now, General Maddox has updated me before this meeting to a couple of things they are cleaning up in Nevada due to the size variances of the mirrors from the original.

"I need everyone to see and understand exactly why I, along with President-elect Slater whom I've asked to join us in this meeting, are pushing so hard on this venture. I believe Dabar could very well turn the economic state of our country back to where people actually begin to believe in the American dream again."

A hand went up from the other end of the table and Secretary of State Sheila Hicks spoke up. "Mr. President, everyone here may not be up to speed on the Dabar expedition. Would you mind elaborating as much as you can?"

"Sure, Sheila," the President said. "That's probably a good idea. General Maddox?"

"Certainly Sir," the General said respectfully before turning everyone's attention to the flat screen monitors mounted on each wall. He clicked a button on a remote control and the screens lit up with still images of the Dabarian countryside. "As you can see, without me telling you otherwise, one would be led to believe that they were in the countryside of some American town. This, however, is a photo taken on the planet Dabar. The planet appears to be for the most part undeveloped. Our current contact there has stated that the planet is very undeveloped and rich in resources. The climate and other things are very much identical to ours and interesting enough, their oxygen mixture is a lot purer and richer than ours. One thing we have not seen is any bodies of water, but this could be due to the fact that we have done no aerial reconnaissance."

The General clicked his remote again and the screens switched over to a shot of the Sanctuary. "We have been given pretty much free reign of the land, and if people are encouraged to go there and actually live there, there will be a need for everything from food, to restaurants and utilities."

"Think of it," said President Duffy. "Everything you need on Earth, as the population grows, you will need on Dabar. This has the potential to create a boom in building and production of every kind."

"But Sir, is it safe?" asked another one of the Joint Chiefs.

"Mr. President, if I may?" interjected Maddox.

"Certainly, General," said the President.

"I can vouch for it being safe," General Maddox assured them all. "I have been there personally on the planet. There is a unique hazard as we have mentioned, but as we speak our team is on the planet being trained on how to make that hazard a non-issue."

"Look," President Duffy added. "We have a ways to go before we begin allowing ordinary citizens on the planet. We want to take every precaution we can to ensure the safety of all of our citizens. However, our plan is to be completely up front with the American people and let them know there are risks and then let them decide. With the way things are happening in today's world, it's a risk to step out of your front door every morning. This is no different, but the choice is up to them."

The head of the NSA leaned forward. "Sir, do we plan to put things in place that would release the government of liability in the event of an incident?"

"I'm glad you asked," President Duffy said. "We have just been working on a plan for that and I want you to work with the General's counsel to come up with a release of liability document that we can use in the applications we will be offering to our citizens along with mandatory health screenings. But for all practical purposes this is an 'at your own risk' deal. It's our intention to be very up front with everyone involved."

"Once we have developed and finalized the training program," Maddox offered, "if people follow the instructions in the training, the risk should be minimal at best."

"Listen," the President said, closing his dossier, "we will convene again next week. Right now everyone have a great Thanksgiving and enjoy your families. Let's get out of here.

The room filled with the buzzing chatter of its occupants as the men and women in the meeting began to get up and leave the room. General Maddox stayed behind however. "Mr. President?" he asked.

"Yes, Jack."

"Sir, Mr. Jergensen should be returning with his team as we speak. Ill forward an update to you as soon as we debrief."

"Great!" said the President. "Be sure Bill and the NSA get a copy as well."

"Yes, Sir." They shook hands and soon the President was gone. As he left, Maddox turned to the only other person left in the room, his assistant. "Make sure those debrief reports are edited before they're issued to the other agencies," he ordered.

His assistant nodded. "Understood, Sir."

Troyconics Astrophysics — Columbus, Ohio, U.S.A.
Earth Date: November 25th, 2014 — 6:28pm EST

Back in Jerry's lab, the original Troyconics insertion team were busy prepping the Dimensional Mirror for operation. It was time to receive Jerry and the rest of the trainees from Dabar. They were coming home. After General Maddox had replaced some of the key technicians with his own men however, they were having some trouble working as the well-oiled machine they once were.

"Tim," Cynthia Raddick complained as she watched one of Maddox's replacements struggle with his task. "The mirror should have been up five minutes ago. What's the problem?"

"I'll have it up in a minute," said Specialist Tim Delchick, who had replaced Ian. "Just having a bit of trouble with the power coupling. There it is." He adjusted a few settings on his console and nodded. "Power is at one hundred percent."

Finally, the team powered up the Dimensional Mirror and watched with wonder as the portal opened. They had seen the mirror work a dozen times by now, but it was still an incredible sight. Moments later, Jerry stepped through the portal with his wife and daughter close behind. After them, the rest of the trainees poured into the laboratory.

"Alright Mario," Cynthia said. "Can you help me get a contagion check?"

As the two of them scanned the trainees with medical wands, and confiscated their bags for further inspection, Jerry looked at Cynthia concerned. "You guys were late getting the portal open," he said. "Was there a problem?"

"Yeah," she told him. "The new guy had a bit of trouble getting the couplings to full power."

"New guy?" Jerry asked, looking around. "Where's Ian? Where's Rajin?"

"They're in Nevada."

Jerry spotted Calvin walking in through the blast doors and broke away from Cynthia to confront him. "Calvin, what's going on? You ship my power expert and gateway expert out while we're off planet?"

"It wasn't my call," Calvin said. "It was Maddox with orders from the President."

"Why does it seem like every time I leave something else happens?"

Calvin shrugged, and then brightened as he spotted Candy and Lexi walking over to him. "Hi Candy. Hi Lexi."

"Hi Uncle Calvin!"

"Welcome back."

Candy gave the family friend a hug and then leaned over to her husband. "Jerry, I'll meet you in the dormitory."

"Okay, Honey," Jerry told her, then grabbed Calvin by the shoulder and motioned him over to a nearby computer console.

"You know the President ordered new mirrors made," Calvin told Jerry. "But what they neglected to tell us was that they made the mirrors larger... Much larger."

"For what?" Jerry wanted to know.

"Oh, tractor trailers, transport trucks... Who knows? Anyway, they're testing the new mirrors, but of course bigger mirrors require more power. So, they needed Ian and Rajin to help make adjustments and rerun power requirements. They also took Sara."

"Choremo? For what? Where's Maddox?"

"Jerry, just calm down and stop," Calvin said. "This is getting further and further from us."

Just then however, as if in answer to Jerry's last question, they saw General Maddox walk into the lab. He headed straight for Sergeant Barclay and the two began to talk in hushed voices. Fuming, Jerry walked over to them, Calvin hot on his heels.

"Jergensen," the General said when he saw Jerry coming. "Glad you're back."

"Yeah," Jerry said unamused. "I'm glad we made it back too. It gives me a lot of comfort to know that in case something happened you put a rookie on one of the most vital aspects of the mirror. This couldn't have waited?"

"Not unless you're giving the President orders, and last time I looked you weren't ranked that high, so get over it."

"I want to talk to the President," Jerry declared.

Maddox laughed at the idea. "For what?"

"I don't believe he knows the whole story. Especially on constructing larger mirrors amongst other things."

"Jergensen, I'm sure the President is on his way to Camp David to relax for turkey just like you will, after giving me a full debrief."

"I'll debrief after I talk to the President. I'm not going to let you, or him for that matter, put my life and the life of my family and these other people in jeopardy for some hidden secret agenda!"

"There is no agenda, Jergensen," the General insisted. "The President is doing what he believes is best for the American people. Now, I told you he's unavailable."

"Great then," Jerry said with a shrug. "My debrief will be unavailable until after the holiday. Happy Thanksgiving General." And with that, he walked out of the lab.

6:53pm EST

Still angry from his run-in with Maddox, Jerry made his way up to the dormitories on the third floor, and entered the private room that he shared with Candy. She turned to him as he walked in and smiled. "So? How'd it go?"

Jerry just shook his head. "Let's not."

"Okay. Can we agree to leave the bad attitude here though?"

"Sure," he nodded as he sat down on the bed and began to undress.

"Good," Candy said. "One more question and then I'll let you shower. Lexi wants to know if Kendra and her Mom can come over tomorrow for Thanksgiving. Their refrigerator went out while she was at work and most of their food spoiled."

"Oh baby, we just got back and I really just wanted to—"

"Jerry, don't be selfish," she scolded him before he could even finish.

He nodded with a sigh. "Sure Honey. You're right."

She leaned over and kissed him before letting him get up. As he made his way across the room to the bathroom to take his shower, Candy gave his nude buttocks a hearty smack. When she opened the door to leave though, the woman who happened to be walking by peeked inside and got a full view of Jerry. It was Ms. Jankowski.

"Oh my!" the old woman cried out.

Candy slammed the door shut quickly and burst out laughing. "Sorry!" she called through the closed door.

"Sorry, Ms. Jankowski!" Jerry added.

6:58pm EST

Sergeant Barclay had just finished showering and changing into a fresh uniform when he heard a knock at his dormitory door. Crossing the room, he opened it to find Lisa Hansley standing in the hallway holding the small audio recorder he had given her.

"Hey, you remembered," he said, taking the recorder from her. "Thanks."

"No problem," she told him. "Check's in the mail, right?"

"Right. If you've got direct deposit it should be in your bank account."

"Great. Well I'm glad you're okay."

Barclay looked at her, confused. "What do you mean?"

"Nothing," Lisa said. "It's just… When we were going through the mirror, I thought you were right behind me. You just took a while to come through and I was kind of concerned something happened."

"Oh," was all he said.

"But I guess it was all for nought. So… Thanks again, and have a good Thanksgiving."

"You too, Lisa," he told her with a smile. They stood there in silence for a moment, looking at one another before Lisa broke away and walked off down the hall. Sergeant Barclay watched her go, then closed the door and took out the audio recorder.

Exhausted from their flight back from Nevada, Ian and Rajin entered the Troyconics facilities and made their way through the security checkpoint. Once they were cleared and given back their bags, they found General Maddox and Sergeant Barclay waiting for them.

"Welcome back," the General told them, handing each of them an envelope.

"What's this?" asked Ian.

"Just a little way of saying thanks from the U.S. Government and the President," Maddox said with a wink.

They opened up the envelopes and peeked inside. Each one contained a government check made out to them in the amount of three thousand dollars. "Wow," Ian said. "Thanks!"

"No," said Maddox, shaking his head. "Thank you. Now you guys get out of here and have a happy Thanksgiving."

They turned to leave, but as Rajin pulled away Maddox put a hand on Ian's shoulder, stopping him. "Ian. Can I talk to you for a moment?"

"Sure," he said. "What's up?"

"Hey, while you guys were out we discovered a way to get more power by using ceramic couplings instead of the other ones. We were supposed to get results to the NSA and the DOD first thing this morning but we had a couple of hiccups. Now I'm not going to hold you up, especially since you just got back in town, but Field Specialist Delchick is coming back in later tonight to help us run those comparative tests."

"I could stay and help," Ian offered with thoughts of more envelopes and checks in the future.

"No," Maddox said. "No, from what I understand it takes about six hours. Go home and enjoy your family. If you could though, just fire up the mirror. It takes a while to get things up to full power and we can save that time and have everything ready to go when Delchick gets here. Would you mind doing that?"

Ian thought about it and was hesitant. But he also realized for all practical purposes General Maddox had the authority to ask him to do so. "Sure," he finally said.

In the depths of Jerry's laboratory, Ian helped the General and his men prep and then power up the Dimensional Mirror. Soon the giant mirror blades were whirring at top speed and the laser light flooded the center, expanding until it formed the shimmering gateway to Dabar.

Ian turned to the General. "Just make sure if Delchick doesn't show up to call me. I'll come back up and power everything back down. If you don't power down the system correctly you can blow the field generator."

"Understood," Maddox nodded. "Happy Thanksgiving."

"Happy Thanksgiving to you too, General," Ian said before collecting his belongings and leaving through the gap in the blast doors. Once he was gone, Sergeant Barclay stepped out from behind a cluster of server towers. In one hand, he held a heavy-looking case of hard plastic.

"The rest is up to you," General Maddox told him.

Barclay nodded and laid the case on the floor, popping the latches and lifting the lid. Inside, ensconced in protective foam, was a miniature plane probe. The Sergeant took the remote control from another compartment in the case and switched the device on. Maneuvering it deftly with his thumbs, Barclay brought the drone up to eye level, and then guided it toward the Dimensional Mirror. Barclay and the General stepped just outside the insertion point as well.

"Great," the General said as the device disappeared through the portal. "Let's see what we can see…"

Watching on the screens built into the underside of the case's lid, Barclay and General Maddox tried to see what the drone's cameras were picking up. Unfortunately, as the device was technically on another planet, the images were delayed, vague and fuzzy.

"Alright, we have a picture," Barclay said after making some adjustments, "but it is faint and fading in and out."

"Can you boost the signal?" the General asked.

"No. You have to remember, there are no transmission towers that we know of on Dabar. So what we're seeing is a delayed signal being relayed through the mirror."

Soon, they saw the Sanctuary of Us come into view, looking almost small as the drone flew over it, high above. "Should I circle the Sanctuary?" Barclay asked.

"No," Maddox told him. "Stay away from it. I don't want to alert any of them."

"What makes you think they're not already aware?"

Barclay kept propelling the drone forward with his hand-held control console, soaring high above the Dabar countryside. Soon, a new building appeared in the camera's periphery.

"Is that what I think it is?" said Maddox.

Chuckling, Barclay shook his head. "Yes sir. I think it's a McDonald's. I'll try to go in closer." He nudged a joystick with his thumb, sending the drone toward the structure, but as soon as it got close, the image cut out and was replaced with a frenzy of crackling static. "Sir, we've lost the signal and the plane is not responding," Barclay announced.

Maddox barely heard him. He cupped his chin in his hand, pensive and somewhat disturbed. "A McDonald's…" he muttered. "But how…?"

Jergensen Residence — Westerville, Ohio, U.S.A.
Earth Date: November 27th, 2014 — 3:43pm EST

For the first time in a long time, the Jergensen driveway was filled with cars. The inside of the house was loud with the talk and laughter of family and guests, who milled about the house, almost swimming in the potent aromas of turkey, stuffing, gravy and a plethora of other delicious holiday delicacies. In the kitchen, Candy and Kendra's mother put the finishing touches on the turkey, transferring it from the oven to a china serving platter.

"Seriously Barbara," Candy was saying, "you really didn't have to buy or make anything. We wanted to have you over as a guest."

Barbara laughed. "Listen, my momma may be six hundred miles away, but she would miraculously appear and whip my butt if I even attempted to come to someone's house and eat without bringing anything or attempting to help with the cooking. Plus, to be honest I had selfish reasons. You know how it is on Thanksgiving. Sometimes you just want some of your own cooking. Nothing against your skills though."

Now it was Candy's turn to laugh. "Trust me. I am not offended. Whatever you don't like, don't eat."

"You know I'm so glad you said that, because I hate being fake. Especially when I'm hungry. You know what I mean?"

"Yeah!" Candy exclaimed. "Then you're fake, hungry and mad too!"

They both burst out laughing again as Jerry walked into the dining room. His eyes bulged as he surveyed the enormous spread being prepped for dinner. "Wow! Are we feeding the army?"

"Jerry, no," Candy waved him away. "Barbara just desired to contribute so she added a few of her recipes to the dinner list."

"Oh, okay!"

"Is dinner done?" Lexi asked as she and Kendra popped their heads into the kitchen. "I'm getting hungry."

"Sure is," Candy said, rinsing the turkey grease from her hands. "If you could put out the drinks young lady, we can all eat."

"Ohh!" Barbara cried out. "I almost forgot." She went over to where she had put her large purse and produced two bottles in brown paper sacks. "Now I don't drink, but I got to have me a bottle of Martinelli's every holiday."

"That's great!" Jerry said, taking the bottles from her. "We love Martinelli's too."

"Well it's a party then," Barbara smiled.

Together, they brought all of the food and drinks out to the table, where Lexi and Kendra had already arranged everyone's place settings. Once everything was set, and all of the glasses were filled, the five of them sat around the table. Holding hands and bowing their heads, they began to say grace.

Easton Shopping Center —Columbus, Ohio, U.S.A.
Earth Date: December 1st, 2014 — 2:03pm EST

Enjoying their leave from the tutelage of the Us, Samantha Morgan sat quietly on a bench, watching her sons play in the fountain at the center outside of the mall. The water shot up from the holes in the flat installation, squirting at random intervals and making the children playing in it squeal with glee.

She loved to watch them play, and she smiled as they ran amok in the spray of the fountain. Soon however, her stomach grumbled and she looked at her watch. "Okay guys!" she called. "Let's go. We need to get something to eat and make a few stops."

"Aw, Mom!" Max bellowed. "Come back and pick us up."

"Yeah!" Cried Mack.

"I don't think so, young man. Let's go." She walked over and took them each by the hand, leading them out of the fountain and back to the parking lot. "Alright guys," she said as they approached the family mini-van. "Dry off good and get your clothes on so we can eat. I'm staving."

"Mom!" Max gasped. "You're not starving. Do you know what the word 'starving' means?"

"No," Samantha smiled, "but I am sure you're going to tell me."

"It means to suffer severely and die from hunger."

"Okay..." she said while the boys finished drying off. "Well maybe I'm just real hungry."

"Yeah," said Mack. "Me too." They laughed and talked together as they took the short walk back to their car and the boys climbed into the back of the van. Samantha got in the driver's seat and cranked the ignition and headed for the closest Red Robin.

Quilsec Astrophysics — London, U.K.
Earth Date: December 2nd, 2014 — 1:41pm GMT

As Bernard Maychoff's Bentley brought him closer and closer to the main entrance of the Quilsec Astrophysics facilities, he felt a *buzzing* in his breast pocket and took out his cell phone to find that the caller ID had been restricted. He answered it and put the phone to his ear without speaking.

"*Bernard?*" a familiar voice said after a moment. "*This is Patricia.*"

Maychoff smiled. "Madame President-elect. What an unexpected surprise. Did you have a good Thanksgiving?"

"*It was great,*" Slater told him. "*I ate too much. But listen, I have to go into a meeting soon so I don't have much time, but are you doing business with a company out of Morocco named Driess Metals and Dyes?*"

Chewing his lip in thought, Maychoff finally nodded. "Yes, I believe we are. The name does sound familiar."

"*Okay, well you may want to find another vendor. They're under investigation for terrorist funding and that link is what put you on the radar. Can you clean this up pretty quick? Things are moving here a lot faster than I expected in regards to Dabar.*"

"Consider it done," Maychoff told her. "Thank you."

"*Thank you, Bernard. I have to go now. We'll be in touch*" she said before hanging up.

Maychoff looked out the window. The Quilsec building loomed closer and soon they were passing through the security gate at the North entrance. Maychoff opened his phone again and dialed a number.

"*Quilsec Astrophysics,*" a pleasant voice answered. "*You've reached Mr. Maychoff's office.*"

"Mary, this is Bernard."

"*Good afternoon, Mr. Maychoff. How can I help you?*"

"I need you to call Human Resources and have them send Brian Ridley to my office now. I will be there in about five minutes."

"*Yes sir. I'll have him wait for you in the lobby.*"

"Have him wait in my office," Maychoff directed. "Ask him to have a seat in there."

"*No problem.*"

He hung up the phone just as the Bentley rolled to a stop. When his driver opened the door for him, Maychoff stepped out and entered the large building. Taking his private elevator, he went up to the executive floor and pulled out his cell phone again.

"*Quilsec Vendor Relations, this is Benjamin Hill,*" a man's voice greeted him.

"Benjamin, this is Bernard Maychoff," he said as he strutted down the hallway toward his penthouse office.

"*Oh, Mr. Maychoff. Hello! How are you?*"

"Good. Listen, I need you to take care of something for me immediately."

"*Sure.*"

"Please sever all ties with Driess Metal and Dyes immediately. Also check for any ties with any of our other companies and or affiliates and discontinue those relations as well. Any payments due to them please forward those to our legal department."

"*I'll take care of it right away, Sir. Was there anything else?*"

Maychoff passed the double doors to his office and continued on to another, nondescript door. Opening it, he entered a small antechamber that, while invisible from his office, was connected to it by a door disguised as a section of drywall. Inside this antechamber he could see Brian Ridley on a series of surveillance monitors that showed the main office from several angles. Ridley sat there alone in front of Maychoff's humongous desk, looking slightly nervous.

Maychoff smiled. "That will be all, Benjamin. Thank you."

Then he hung up the phone, slid it into his pocket and went through the secret door, into his main office. Surprised by his sudden arrival, Brian Ridley jerked up in his chair, caught off guard. "Mr. Maychoff, it's uh, a pleasure to finally get to meet you."

"Really?" Maychoff asked.

"Yes sir. I'm a big fan."

"Of?"

"Just of your continued dominance and acquisitions track in the astrophysics field," Ridley stammered.

Maychoff walked past him to the standing bar in front of the window and poured himself a beverage. "Would you like a drink?" he asked Ridley.

"No sir. Not while I'm working."

That's when Maychoff turned suddenly, eyes now sharp and blazing. "And who are you working for…right now?"

Ridley's face went slightly red, but he attempted to keep his composure. "Well sir, uh, ultimately that would be you."

Maychoff nodded intently as he walked back over behind the desk and stood, taking a sip of his drink while peering at Ridley over the rim of his glass. "Mr. Ridley, I truly applaud your job performance and work ethic. They are both unquestionable."

"Well thank you, Sir—"

"Your integrity however…is not."

Ridley said nothing. He sat there, like a deer in headlights, waiting for the hammer to drop.

"It has come to our attention," Maychoff went on, "that you have been doing a bit of *moonlighting* for—shall we say—your *other employer*… And in light of that, your services will no longer be needed."

Just as he spoke Ridley's firing, there was a soft knock at the door and Mary, Maychoff's assistant, entered the office holding a medium sized file box. Behind her stood two looming security guards.

"Mary has been kind enough to gather your things," Maychoff told the stunned Ridley, "and these gentlemen will see you out. I'm sure you'll have no trouble in acquiring gainful employment elsewhere. Good day."

Then Maychoff lowered himself into his chair and spun around, showing Ridley his back. He listened as the blindsided and disoriented Ridley processed in silence what he had just heard and finally got up, taking the box from Mary.

He looked back at Maychoff one last time, trying to think of something to say, but eventually he left without comment and Maychoff was left alone in his office.

2:17pm GMT

In the employee-parking garage of Quilsec Astrophysics, Brian Ridley sat in his car. He felt an odd mixture of guilt, anger and terror. He felt guilty for betraying Maychoff and being caught. He felt angry at being let go in such a humiliating fashion. And he felt scared about what he would do next.

Trying not to freak out, he rummaged around in the file box Mary had given him until he found his phone—the burner phone he kept in his desk. He flipped it open and dialed a number.

"*Bill Tish,*" the Director said upon picking up.

"Director Tish. This is Brian Ridley."

"*Mr. Ridley. How are you?*"

"I'm fired Sir," he reported, ashamed. "My cover's been blown at Quilsec."

"*We're aware of it,*" Tish told him.

"Well what happened?"

"*We don't know yet. There were some inquiries made at a very high level. We believe that's how your cover was blown. Either way it's inconsequential. Maychoff was not engaging in anything illegal. He just happened to be doing business with a company that is. His director of vendor relations severed all ties with those companies just a few minutes ago, and froze all the remaining payments due to those firms.*"

"Do you think that's a coincidence?" Ridley asked.

"*No. Someone tipped Maychoff off and brought it to his attention and Maychoff is shrewd and prudent. He's not going to do anything to jeopardize his citizenship, or his business dealings.*"

"So he's no longer under investigation?"

"*We'll keep an eye on him,*" the Director assured him, "*but officially? No. Come in and I'll brief you on your next assignment.*"

Ridley nodded, happy at the prospect of further work. "I'll see you in the morning."

He hung up the phone and threw it back in the box. Then, thinking of something, he dug it out again and scrolled through his contacts, looking for another number. Finally he found the name he was looking for: *General Maddox.*

Troyconics Astrophysics — Columbus, Ohio, U.S.A.
Earth Date: December 2nd, 2014 — 9:17am EST

In the conference room at Troyconics, four men sat around the table, ready for Jerry's debriefing. Split into pairs, Jerry and Calvin sat on one side of the table. Across from them sat General Maddox and Sergeant Barclay. The General and Jerry were glaring at each other across the tabletop, neither trying to hide their dislike.

"Let me state for the record," Maddox grumbled, "that from here on out, *all debriefs* will take place immediately following the mission with the exception of family or medical emergencies."

Jerry said nothing.

"Is that clear?" the General growled through gritted teeth.

The question hung in the air as if doing a tightrope walk on the tightly wound tension in the room. Before Jerry or Calvin could answer, however, the General's phone began to ring. He took it out and looked at the screen. "Excuse me I need to take this."

The General stood up and walked out into the hall, thumbing the button to answer the call and putting the receiver to his ear. "This is General Maddox."

"*General, this is Brian Ridley,*" a familiar voice came through the phone's speaker.

"Mr. Ridley. How can I help you?"

"*General, I was just moved from an undercover assignment that I'd been on for years. I don't know exactly what happened, but I was just wondering whether or not you shared any portion of our meeting concerning Maychoff with anyone.*"

"Mr. Ridley," Maddox said into the microphone, "that information is classified and, to be quite frank, I don't appreciate the insinuation."

"*Well with all due respect, General, I'm a little flustered myself and I'm just trying o get to the bottom of how Maychoff discovered I was working for someone else. No offense was intended.*"

Maddox sighed. "None taken. Just know it wasn't from here."

"*Thanks, General,*" Ridley said and then the line went dead.

Frowning, Maddox stuffed the phone into his pocket and stood there in the hall, thinking. When he had sufficiently collected his thoughts, he opened the door to the conference room and entered once more.

"Thanks for your patience," he said. "Now, moving right along. First off, besides this new resort that I've heard about from Sergeant Barclay, are there any additional buildings or facilities that we need to know about?"

"Not to my knowledge," Jerry said. "Did you see anything else Sergeant Barclay?"

"Not while we were there," said Barclay.

"General, if the Us decide to add a room or additional building they aren't applying for a permit from us."

"So what are you saying?"

"I'm saying that whatever pops up between or during our visits is not under our control. And I don't know that it's any of our business."

The General and the Sergeant turned to each other, sharing a knowing glance that Calvin couldn't help but notice. "Are we missing something here?" he asked.

"What do you mean?" said the General defensively.

"Well it appears that you are insinuating that more has taken place on the planet than mentioned. However, we haven't even begun the debrief so why would you be inquiring about it. Unless of course you know something we don't."

"I wasn't insinuating anything," Maddox said with a scowl. "I simply want to know what facilities are available to our people. Jergensen, let's just move on and get your full report."

"Sure," Jerry said snidely. "Oh, by the way did Barclay tell you about the Hummers?"

"The Hummers?" the General inquired surprised.

Craden's Restaurant — Clintonville, Ohio, U.S.A.
Earth Date: December 2nd, 2014 — 11:14am EST

Bustling with the sound of clinking glasses and silverware, the restaurant was packed with families enjoying brunch. In the rear of the restaurant a single female sat alone, lost in thought.

Her meditation was interrupted as someone walked up behind her. "Is this seat taken?" they asked in a familiar voice.

Field Specialist Tamyra Ann turned in her seat to see Jeremy Moore standing over her. Looking handsome and stylish as always, her boyfriend—or was he an ex-boyfriend now?—looked down at her with his usual winning smile.

"Hey you," Tamyra said, caught of guard and a little confused.

He leaned over and kissed her on the cheek before sliding into the booth across from her. "What's up girl? I missed you."

"Really," she said. "What have you been up to?"

Jeremy shrugged. "Working and waiting for you to come back. Listen, I know we kinda left each other in a heated discussion but I apologize. I overreacted. I know we can work this out."

Tamyra looked into his eyes, wanting to believe him. But she had to shake her head. She had to be strong.

"What's up?" he asked, sensing all was not right. "Are you still mad? Okay, I know this is not what you called me here for. I know you're not trying to break up with me over some pettiness."

"Jeremy, that's just it," she said. "It's petty to you. It was petty to me the first eight times it happened. I mean the temper, the name calling. The immaturity. I just don't think I can take it anymore. I don't think I want to take it anymore."

The cowl of anger slid over his eyes once again like a magic trick. "Where's your fortitude?" he hissed at her. "Where's your mental toughness 'army lady?' You're *weak*! You say you love me, but when a little trouble comes, you run!"

As the volume of his voice climbed higher and higher, more and more people at the surrounding tables began to look over at them, staring. "See, this is exactly why I'm out," Tamyra explained, not caring about the spectators. "You can't get angry without attacking a person's character, and I don't I *won't* put up with a person like that. I know I deserve and can get better."

"Fine!" he spat at her. "Speed on before you get pee'd on…" he got up out of the booth with a huff, attracting the attention of an old man sitting nearby. "Whatchu looking at, Grandpa?" Jeremy said to him. "Mind yo' bitness!"

"Lord, why do I keep attracting boys?" Tamyra said to herself, shaking her head as Jeremy stormed out of the restaurant.

Troyconics Astrophysics — Columbus, Ohio, U.S.A.
Earth Date: December 2nd, 2014— 11:52am EST

Closing the door behind them, Calvin and Jerry left the conference room exhausted from the lengthy debriefing. As they made their way down the hall, Calvin shook his head. "Did you have to bring up the Hummers?" he asked.

"What?" Jerry said. "We have Hummers now. The army has Hummers, now we have Hummers."

"You know if I didn't know better, I would say that the General is jealous."

"Yeah, me too, but of what?"

Just then they saw Ms. Jankowski walking in the other direction, along with Samantha Morgan and her two sons. They all smiled as they converged in the middle of the corridor.

"Samantha," Jerry greeted them. "Max, Mack. How was your Thanksgiving?"

"Great," Samantha said.

"Yeah," exclaimed Max. "Mack ate too much."

"So did you!" Mack shot back.

Samantha laughed. "Keep it moving, guys," she said, ushering them further down the hall toward the dormitories.

"Good morning to you, Jerry," Ms. Jankowski said. "Mr. Madison. How was your Thanksgiving?"

"Ms. Jankowski," Jerry said slightly blushing, still embarrassed over Ms. J seeing him nude. "Good, how about yours?"

"It was awesome," the old woman said with a smile. "I was simply ecstatic to have some real food." Then she leaned forward and spoke quietly to Jerry. "Jerry, you don't have to be embarrassed because I saw you naked. I'm old enough to be your mother, you know?"

The words made Jerry's face flush red even more as he nodded. "Thanks, Ms. Jankowski. I feel much better now."

"Good!" she said before continuing on her way with a mischievous smile. "I'm glad. See you soon. Hopefully clothed this time."

As she walked away from them, Calvin turned to Jerry, stunned. "She saw you naked? Does Candy know about this?"

Jerry nodded, still bewildered and embarrassed. "She was there."

"Really?" Calvin said, his eyebrows raised in a silent request for more details.

"It's a long story. Back to the General."

"Right," Calvin said as they resumed their stroll toward the service elevator that would take them down to the lab. "He and Barclay are up to something. I don't know what."

"The truth will come out," Jerry said. "It always does."

They climbed into the elevator and began their descent to the subterranean laboratory. When the doors slid open, however, they found themselves standing face to face with—

"Ian!" Jerry cried out in pleasant surprise. "You're back. At least I don't have to worry about delays this time."

"Jerry," Ian said, shaking his hand. "How was your Thanksgiving?"

"Good. Yours?"

"It was fine," Ian told them. "Hey, do you know if General Maddox and Sergeant Barclay got their mirror test results finished?"

Calvin and Jerry looked at one another. "What test results?" they said at the exact same time.

"Uh," Ian said, realizing that they hadn't known. "Never mind."

"Ian?" Jerry said.

"Well, we flew back in late Wednesday night and General Maddox and Sergeant Barclay were still here. General Maddox asked me if I could fire the mirror up for him so that Delchick didn't have to wait for it to warm up. He said they were replacing the regular couplings with ceramic couplings because they produced more power and they had them installed but needed to get the test results to the NSA and DOD."

"Oh, okay," Jerry said. "Thanks Ian. I'll make sure they get those."

They let Ian onto the elevator and made their way toward the blast doors and the laboratory behind it. "The truth will come out," Calvin repeated.

Jerry nodded. "Always does."

They were in the lab for fewer than five minutes when General Maddox and Sergeant Barclay appeared themselves. As soon as Jerry saw them coming through the blast doors he started toward them, ready to confront the General, but Calvin reached out and stopped him.

"Jerry, don't," he said. "Wait. When you wait, deception always reveals itself."

"Are you sure?" Jerry asked.

"It worked with you," Calvin smiled.

"Oh, that's low," Jerry said. "That is low. I thought you forgave me." Though he couldn't help grinning slightly himself.

"Alright everybody!" Maddox barked. "Let's get ready for insertion. Jergensen!"

"Yes sir, General," Jerry said giving Calvin a knowing wink.

Maddox walked up to him and handed over two large cardboard boxes and a poster tube. "So here are your Hannibal Lecter masks and the drawings and schematics you asked for during the debriefing. You plan on building something?"

"Maybe," Jerry shrugged. "You have your secrets, General, I have mine. Sergeant Barclay, can you grab one of those boxes for me?"

Jerry took one of the boxes from the General and walked away. Maddox followed him with his eyes and smiled as Sergeant Barclay stepped forward. "Sure Boss," he mumbled.

"Don't worry," Maddox told him. "He won't be boss for much longer. Hell, he won't even be around."

Barclay snickered as he took the other box from his General.

The Sanctuary of Us — Dabar
Date & Time Unknown

Back on the lush landscape of Dabar, the team's Hummers made their way once more up the rise of the valley and over to the castle and resort. As the SUVs parked and the human trainees climbed out of them, Jerry noticed a new building several hundred yards away. A McDonald's.

"Greetings, Jerry," ThreeofUs said as he walked over to the group. "Was your Thanksgiving…thankful?"

"Yes, it was," Jerry told him.

"I sense a hint of deception, Jerry. Are you being truthful?"

"Well," he sighed, "Candy asked to have guests over and I really didn't want to, but I acquiesced—"

"Good, Jerry. Selfishness will destroy life from the inside out."

Jerry nodded. "I'll remember that. What is the deal with the McDonald's? Did we buy a franchise?"

The ThreeofUs shook his head as if the question had been serious. "No. We constructed it shortly after you left from the pictures in Max and Mack's minds. However, we do not have anyone to staff it. We have discussed allowing additional people to come and staff it, but unlike your team they will need to leave each evening when their schedule is completed. Once they have been properly trained we can construct living accommodations for them as well."

"But what about them speaking?" Jerry asked. "Will you create a Grace Asylum for them as well?"

"Grace Asylums are created only out of necessity and, with the exception of the Sanctuary, we have only a limited period of time for Grace in those others areas."

"So eventually the Grace will run out in the resort area?" he asked, suddenly concerned.

"Eventually."

"So what will they do?" Jerry wanted to know.

The ThreeofUs pointed over to one of the Hummers, where Sergeant

Barclay was busy unloading the cardboard boxes that they had taken from the General. "We will use your chastity muzzles."

Then the being went over to the boxes, opened one deftly with a single finger, and took out one of the muzzles. Holding it to his mouth, he began to murmur and whisper into it, making minute alterations with nothing but his words. When he was finished, he simply said "copy," and the rest of the muzzles in the boxes conformed to the same specifications.

Jerry watched all this, amazed at the ease with which the ThreeofUs had bent the objects to his will.

"Come everyone," ThreeofUs said when the Hummers were unloaded. "We have much to do today."

Date & Time Unknown

Later, after they had resettled in their rooms, the trainees met back up in the classroom of the Sanctuary. The ThreeofUs stood in front of the blackboard as usual, waiting for them to find their seats and give him their attention.

"Once again, one of the first and most important things everyone who comes here must learn," he began once everyone was in attendance, "is to unlearn what you have learned. So much of natural Earth life is based on limits. What one can do and what one cannot do.

"A person's life and subconscious begins to be trained and molded into this shape by the words they hear. This type of behavior automatically limits creativity, imagination and freedom, and therefore limits life. Mack can you come up to the front?"

Mack looked over to his mother, who nodded. Then he slid out of his seat and made his way to join the ThreeofUs at the front of the class. "When an adult is told something," the ThreeofUs went on, "what is the first thing they do, inwardly?"

"They analyze it," Jerry answered. "Question whether it is true or whether or not they can do what was told to them or not."

"Exactly," ThreeofUs said, pleased. "They analyze it. One of the primary reasons they do this is because they have been taught to not trust. Since they have been lied to, they have learned to not trust what people say. Now, what do children do when they are told to do something or when they are told something is going to be done for them?"

"They just believe it!" Ms. Jankowski said, suddenly understanding. "Especially if they're younger."

"True again," said ThreeofUs. "Now here we have young Mack. Mack has great power."

"I do?" asked Mack with much surprise.

"Yes, you do! I will say it again: Mack has great power. He has the ability to perform on many levels here on Dabar. However, his *dominant skill set* with words is in being a *mover*."

Mack looked up at the alien being with the teenage face, his eyes full of wonder and intrigue. "I am?"

"Yes, Mack. You are."

"Okay! What can I move right now?"

The ThreeofUs looked down at him, as if his curiosity was piqued by the simplicity of the child's trust. "The pencil on your desk," he said. "Tell it to come to you."

"You mean…talk to it?" Mack asked, confused. "Like my dog?"

"Exactly. Yes."

Mack shrugged and turned to face his empty desk twenty feet away. "Pencil," he said. ThreeofUs interrupted Mack. "Mack, We need you to be specific, otherwise it is possible that every pencil in the room will move."

"Really!" Mack said, even more excited. ThreeofUs nodded. "Say *my pencil*."

Mack's face maintained all the excitement and faith he had when he first heard of his ability. "My pencil, come to me." "Move!"

At first nothing happened. Then, as if being tugged by some invisible fishing line, the pencil rolled forward along the desktop, heading straight for Mack. After rolling only half a foot it stopped again.

"Wow!" Mack gasped.

"Now," said ThreeofUs, "put your hand at the end of your desk and tell the pencil to come into your hand."

Mack walked over to the desk and stretched out his arm, his hand held out palm up. "Pencil! Come into my hand!"

Again the pencil rolled forward, this time tipping over the edge of the desk and landing perfectly in the boy's grasp. "Cool!" he marveled.

"That is cool," his mother agreed.

"Although most of you have seen this done twice," ThreeofUs told the rest of the class, "many of you will still doubt and believe that the pencil just happened to roll on its own. This is the type of unbelief that you must unlearn. You have been taught not to trust. Your reasons vary. Some of you were promised toys by parents or relatives and the promises were never kept. Beware! It is still affecting you today!"

The ThreeofUs turned his gaze to Ms. Jankowski, who was busy wiping a tear from her cheek. Then ThreeofUs turned to Candy. "Others were told things that were hurtful and discouraging…" Candy turned away from the beings stare, able to look only at the floor as Major Blake was behind her.

"And many of you were told you were loved but the time and attention to back up those statements did not accompany the words," ThreeofUs went on, this time looking intently at Sergeant Barclay, Tamyra Ann and Lisa Hansley.

"The truth may hurt," ThreeofUs continued, "but it always heals. Lies may seem to soothe, but they always kill. What is it that they kill, you may ask? Trust. Once trust is killed the result is always no faith, no trust in another person's word, and eventually no faith, no trust in one's *own* words. Those of you who are struggling to believe, this is what has caused this. This is what you must unlearn."

The classroom sat in silence as the adults pondered the ThreeofUs words while remembering the lies of their past. Both the lies they were told, and the ones they told themselves. They sat there for seconds that seemed like moments, meditating on the concepts and truths that the ThreeofUs had just presented. The silence was only broken when Mack yelled—

"Dictionary! Move!" and the dictionary moved forward slightly, untouched.

The class gasped at the trick and the ThreeofUs stared at the boy. "Mack please, only the pencil right now."

"Okay," Mack said. "But how come the book only moved a little bit?"

"The power in your words and the force you use to do things is like a muscle," the ThreeofUs explained. "The most important thing involved is you actually believe what you say. The next thing is that you develop and strengthen your muscle by practicing. The dictionary is larger, heavier than the pencil, and therefore requires more strength."

Then the ThreeofUs leaned down, close to Mack and whispered to him. "But always remember this: you must control yourself. Just because you can do something does not mean that you should."

Mack nodded absorbing the words of wisdom as the ThreeofUs turned back to the others. "Good," he said. "Class is dismissed. Your homework is meditating on what we have just told you."

As the trainees began to gather their notebooks and dictionaries, the ThreeofUs stepped over to Jerry. "Jerry," he said, "May we speak with you briefly?"

"Sure. What's up?"

"Do you remember when upon your first visit how we denied Sergeant Barclay access to the Sanctuary?"

"Yeah," Jerry nodded. "I was actually surprised when you allowed him to return and be a part of the team."

"Be aware, Jerry. There is nothing we do without purpose, and the purpose for our apparent change of heart is now made manifest. We have foreseen, through Sergeant Barclay, that General Maddox and someone we cannot yet see will remove you from the project and this planet."

"What do you mean?" Jerry asked, shocked. "You said *will*. Not that they're planning to. What have you seen?"

The ThreeofUs shook its head. "We divulge this information to you after much deliberation with AllofUs. The use of the word 'will' is deliberate, intentional, and already in progress. They believe you are a threat to their ultimate goals for this planet."

"Which are?" Jerry pressed.

"We have told you as much as we can," the ThreeofUs said. "Simply know it is inevitable and only a matter of *your time*. Do not make them aware that you know. Jerry, be angry now so that you will not display anger later. They believe they have the upper hand however, there are more that be with Us than be with them. Are you familiar with the passage of scripture where Jesus washes His disciples feet in John, chapter thirteen?"

"Yes, I know it," Jerry said, unsure of what the ThreeofUs was getting at.

"After the feet washing, Jesus and his disciples sat down to eat and John asked Jesus which one of them was going to betray Him. Jesus told John the one to whom He gives the bread to after he dips it in the sop. Do you remember what Jesus said to Judas after doing this?"

Jerry nodded. "He said 'what thou doest, do it quickly.'"

The ThreeofUs turned away from him. "What they do, Jerry, they will do it quickly." Then he walked away from Jerry, calling back to him over his shoulder. "Be at peace, Jerry. Although I will not see you soon from your perspective, *I will see you*."

And then the ThreeofUs departed.

Chapter 9

Ms. Jankowski's Penthouse — Dabar
Date & Time Unknown

Almost eleven months after Jerry and the ThreeofUs conversation, Dabar
has grown from a small settlement of a dozen trainees to a bustling
population of more than seven million people. On the new, off-planet
colony referred to as The United States of America on Dabar, American
citizens of all races and creeds have begun to build new lives on the parallel
world. Buildings spring up daily and an entire city has been constructed in
just under a year. Many people young and old have quickly adapted to the
unique language barrier of planet, some of which have gone beyond the
native language adjustments needed to survive to learn Hebrew, with which
things can be erected on an advanced level. I am one of those people.

I now understand the youthful appearance of many of the Us. The
planet's very atmosphere combined with the removal of negative and death
related words, has a regenerative effect. As I stare back at myself in the glass
of my window now one year older, I still stand amazed at my own face and
body which appear to have gradually shed twenty or so years during my
time on Dabar. I must say it is a strange thing to watch grey hairs disappear
without hair dye. It's almost like a second chance to be young again, to
have fun again. Experiencing the wisdom of age and the beauty and health
of youth at the same time is almost indescribable. Although I am now
seventy-five, I could easily pass for fifty. Heck, maybe even late forties. I
think to myself, "Jankowski you're blessed!" Looking down on the
blooming metropolis from the balcony of my penthouse apartment,
restaurants, houses and apartment buildings provide the once lush valley
with the look of a major American city. On the outskirts of the sprawl, you
can see new buildings being constructed from nothing but words. By
incorporating the life and natural laws of Dabar even my eyesight has
improved so tremendously I can make out a faraway billboard without the
help of glasses that advertises the services of the Max & Mack Mover
Company. My heart swells with pride for the two boys.

However, the boys and others are not the only ones who have learned new things. With the help of the ThreeofUs, I have developed an app to use on the Apple Watch that now serves as a personal six-second box in addition to the standard boxes located throughout the inhabited part of the planet. Imagine me? An app designer at seventy-five! Here, you really can teach old dog new tricks!

Not far from my apartment building, a new, more condensed and larger Dimensional Mirror and portal remains open, acting as a full-fledged security checkpoint for those leaving or coming to Dabar. People mill in and out of it, ushered along by military personnel. Each new person arriving through the portal is immediately fitted with a chastity muzzle.

These rookie Dabarians are now referred to as Hannibals in light of their *stylish* mask for the next few months as they enter and finish their training.

Sadly, as masses continue their exodus to this new world to partake of its now well-known benefits, many still fall prey daily to its unforgiving law regarding perverted speech. In what should be utopia, involuntary suicide is still the leading cause of death on Dabar.

FedEx has even developed a specialty branch on Dabar, devoted to the removal, storage, and next day delivery of bodies back to their loved ones on Earth. DeadEx I believe it's called. The trucks can be seen everywhere throughout the new city streets, picking up bodies wherever people may have misspoken.

It has been less than a year since the ThreeofUs warned Jerry of what was to come and yet, so much has already taken place. Dabar is now a very different place.

Troyconics Astrophysics — Columbus, Ohio, U.S.A.
Earth Date: December 8th, 2015 — 1:26pm EST

One year older, and sporting quite a few more gray hairs, which had sprouted up due to stress over the past twelve months, Calvin Madison sat behind his desk. He had the news on as he filled out paperwork, multi-tasking as he watched a report on Dabar.

"And today President Slater released additional news on what people are now calling 'The Other Earth,'" said the UNN news anchor. *"For the past six months people all over the United States have been migrating to the planet Dabar in hopes of what some have termed a planetary gold rush. The U.S. has experienced a proverbial influx of cash and more importantly labor and work for many who were only a year ago unemployed.*

"President Slater also introduced the newly appointed Secretary of State for Dabarian Affairs, Richard Killix to replace Jerry Jergensen of Troyconics Astrophysics, who was removed from the planet and the project several months ago. Jergensen is still recognized as the discoverer and pioneer of the parallel planet, and also oversaw all initial travel, training and expeditions on Dabar. Jergensen's sudden removal shocked many and no explanation was given for the dismissal, but rumors persist of friction between Jergensen and other key members of the team overseeing operations on the planet.

"President Slater's plan to expand the Dabar insertion to not only U.S. citizens but other countries as well, through the sale of Dimensional Mirrors to qualifying countries, appears to have actually provided a much needed stimulus in what some financial analysts are now saying will eventually become a surplus to the country. The President had this to say:"

Calvin glanced at the screen just in time to see the screen cut away from the network's lead anchor, and replace him with a shot of the President sitting behind her desk in the Oval Office.

"My fellow Americans," she said calmly, *"we have seen tremendous growth and improvement in terms of our country expanding its footprint to include sovereign provinces on Dabar. The economic boost in terms of labor, production, exports and job creation has been almost staggering. I would be*

remiss if I did not mention the fact that some of our citizens have lost their lives in their attempt to overcome the unique language barrier on Dabar.

"However, we continue to refine our training program and implement every protective measure at our disposal to ensure a safe and mutually beneficial transition to the new planet. As this precious discovery has benefited our beloved country and as a leader in the world today, our decision to allow our many allies and others access to Dabar as well, has proved to be a win-win for all involved.

"We are currently working on plans that will increase manufacturing of Dimensional Mirrors for people of other countries so that this blessing may be shared with as many as possible. In light of the exponential growth and the growing complexities that come with governing an annexation of this magnitude, I have appointed Richard Killix as the U.S. Secretary of State for Dabarian Affairs.

"Richard has at least six months of experience in handling affairs on the planet Dabar in multiple facets. He has worked closely with the existing governing bodies that we have instituted on the planet. He has accepted the task of ensuring that all of our citizens are safe and have an opportunity to be a part of this tremendous experience as well as oversee our other international interest there. Richard will work alongside the U.S. Secretary of State in this parallel position.

"As details develop I, along with my office, will release additional information. We invite those of you who have not allowed yourself to qualify for this unique experience to do so. Those who are skeptical, talk to your friends and co-workers who have already experienced what Dabar has to offer. Thank you so much for your cooperation in making this historic times in our country and the world an even bigger success. God Bless you, and God Bless America."

Then the feed from the Oval Office cut out and the screen returned to the UNN sound stage, where the lead anchor was shuffling the papers on his desk. "You know," he said to his co-anchor, who sat off to the side. "one has to think, if we open the planet up to everyone, won't it become just like earth?"

"That's an interesting thought," the anchor's colleague said, pondering the idea.

Calvin shook his head, staring at the TV. "Hopefully not," he said. "Hopefully not…"

Biomedical Research Tower — Columbus, Ohio, U.S.A.
Earth Date: December 9th, 2015 — 10:56am EST

Secret Service agents swarmed about, creating a safe passage for President Slater as she left the Ohio State University auditorium and made her way to the idling motorcade outside. Along side her, the newly appointed Richard Killix followed. His assistant, Alexa Thompson hung back a bit, careful not to intrude or eavesdrop, but ready to make herself useful at a moment's notice.

"You know, Richard," President Slater told the new Secretary of State of Dabarian Affairs, "I must tell you, I am surprised—No, perhaps shocked is a better word—at the rate of growth and expansion taking place on Dabar."

"Trust me, madam President," Killix said. "The pictures do not do it justice. To actually watch the process is like watching a movie. You must visit sometime."

"I said the same thing," the President chuckled, stopping just short of her limousine to continue the conversation. "However, the head of the Secret Service feels they have a complicated enough time just protecting me here on Earth. And since they don't know what type of threats can or can't be done on Dabar, I have not yet been cleared for a visit. Isn't that right, John?"

The Secret Service Agent closest to her nodded with professional curtness. "Yes Ma'am."

"Yes I can go?" Slater asked, teasing.

The agent betrayed the slightest grin. "No Ma'am."

"You see?" Slater asked, turning back to Killix. "The President is just a title. They really run everything." Although the day was clear and dry, the brisk December cold made them long for the heat of the presidential limo. They carried their conversation inside the vehicle.

"Well," said Killix settling into the comfortable seats, "we'll have to see what we can do about that. Who knows, by seeing the Commander in Chief on Dabar, that may persuade even more to come. Not that there's any shortage of visitors."

"So what's the main attraction, Richard?"

"Jobs!" he laughed. "People are either working or are in training to learn how to work on the planet. I mean people need homes. Corporations need buildings and right now there are only a few creative companies to choose from so demand is definitely outweighing supply."

"So how can we speed things up?" the President challenged him.

"Well, it's not that easy. The buildings actually go up really fast. Normally a large office building can go up in about seventeen minutes."

"Seventeen minutes?" Slater said, surprised. "I don't know if I want to be in a building that was put up in seventeen minutes. I'm sure you guys have made sure it's safe, but how does that work?"

Killix laughed heartily. "How much time do you have?"

"I'll make time," she told him.

"I've got a couple of hours before my meeting with General Maddox over at Troyconics," Killix told her.

"Great!" she said. "You like pizza?"

"Who doesn't?"

"Good. I heard Adriatico's has the best pizza in town and they are right around the corner. We can talk while they do security clearances." Then she buzzed the Secret Service agent. "John, Adriatico's, please let them know that we're coming for lunch."

"Yes Ma'am," he said before raising the divider window in the limo.

The Sanctuary of Us — Dabar
Date & Time Unknown

Just as they had a year ago, the ThreeofUs stood at the railing of the Sanctuary's highest terrace. He looked out at the progress the humans had made in their short time on Dabar, studying their expansion with expressionless eyes. After a few moments, he was joined by the OneofUs who stepped out from the darkened archway of the castle and walked over to stand beside him.

"And so it is here," they both said together.

"Yes indeed," said ThreeofUs. "Many are here."

"With more to come," replied OneofUs.

Once again they spoke in unison, their voices intertwining and echoing together. "More indeed," they said.

"If they could but understand..." ThreeofUs then lamented on his own. "...the *vanity* of all they are doing and building..."

"Such insight would require light," OneofUs pointed out. "And so many here are dark."

"Only here for what they can get instead of what they can give," the two beings said together.

"And that is why so few have joined Us," ThreeofUs said.

"Many are called..."

"...but few are chosen." The ThreeofUs nodded.

They stared out at the bustling population below and the buildings that now rose up out of the ground like giant steel stalagmites. After a moment the OneofUs left, disappearing back into the Sanctuary.

The ThreeofUs continued to stare out at the growing civilization before him however. He watched as the humans below milled about, building and working and expanding. At the edge of the human's ever-growing settlement was an untouched valley that separated and marked the property of the Us with a large gate.

As ThreeofUs watched, several humans were passing by the gate, Hannibal's as they were called all wearing chastity muzzles as they headed out into the valley for a hike. The ThreeofUs watched as they climbed across the terrain, exchanging rude gestures with each other since they could not speak. As they approached the gate of the sanctuary, one of the younger humans saw the ThreeofUs watching and lifted his hand. Extending the middle finger of that hand, the human, laughingly shot the ThreeofUs a gesture that was unmistakable.

The ThreeofUs pupil's dilated, shrinking to the size of a pin. "So be it…" the being murmured as it stared intently at the young man that had made the gesture. "he is *katapugon*."

As soon as the words left the ThreeofUs mouth, the young man at the gate below stopped laughing. The human looked up at the ThreeofUs, fear smeared across his face. "No respect for boundaries," the ThreeofUs said to himself. "A picture of things to come…and go."

Then the ThreeofUs turned and left the terrace without another word.

Chapter 10

After taking a few minutes for an obligatory photo-op with the restaurant owner and his wife, President Slater and Secretary Killix had the place to themselves. The Secret Service had swept the place before their arrival, and had emptied out any lingering customers to ensure her safety. Now they took a seat at a booth in the corner and continued their conversation as they waited for their food to arrive.

"So you're saying it takes three to four months just to create the blueprints for a building on Dabar?" President Slater asked her new Dabarian Secretary of State. "They either have some bad architects or the blueprints are made of gold."

Killix laughed. "Neither. And there, they are not called blueprints, they're called *cresigns*. It's short for creative designs. Here things are constructed. There things are…created."

"But three or four months just for…cresigns?"

"Yep!" he told her. "But here's the kicker. When they cresign the building, they *really* cresign the building."

President Slater arched a perfectly styled eyebrow. "What do you mean?"

"When they cresign, they cresign everything in the building. And I mean *everything*. The chairs, the carpet, the faucets, the desks… Everything is in the building when it's created. So once the building goes up, after the structural integrity check and sign off, you can occupy the building in one week."

"Are you serious?" Slater asked, her mouth agape.

Killix nodded, reaching into his briefcase and pulling out his iPad. He swiped it open and, after a few expert taps of the screen, brought up a video. Spinning the tablet around on the table so that the President could see, he sat back and let the video's narrator explain.

"*Your dream,*" a woman's voice came from the iPad, "*starts with words. Words that create pictures. We turn those words into reality. If it's a home, you choose every aspect of what your new home will look like from the wood for your cabinets to the color and texture of your bedroom carpet. If it's your next office building, every desk, every office, every gym, everything will be in place upon creation."*

In the video, footage of construction on Dabar played. The way that buildings and objects seemed to manifest themselves from nothing gave the whole video the feel of a big Hollywood movie. In fact, part of President Slater's mind still wondered if she wasn't merely watching the results of especially convincing CGI or visual effects. But she knew better. This was real, and it was happening on the other planet more frequently every day.

"How you ask? Again, it all starts with words. More specifically, in some cases Hebrew words. Did you know that by using the twenty-two Hebrew letters, a word can be designed to describe anything in the universe? On Dabar, these words can be used to create tangible objects from invisible material called The Substance. *As an architect or designer, you can learn to communicate with* The Substance *and when you do… the possibilities are endless!"*

On the screen of the iPad, the video showed a newlywed couple sitting down at a desk with an architect, selecting designs from a reference book and customizing the cresign for their first house together.

"When you sit down with one of our architects, every detail of your dream or design is meticulously translated into words. Every measurement, every piece of material created from this invisible substance enabling us to bring your dream from plan to manifestation in a fraction of the time it takes on Earth and at a significantly lower cost. Each person involved in the creation is a specialist in their field and all of our creations come with a 100% structural and satisfaction guarantee or we will recreate your dream for free."

The video came to an end and Killix returned his iPad to his briefcase. "It's pretty amazing," he said. "Once they have the cresign, these people who specialize in these words.. these Hebrew words… literally *speak* your house, skyscraper, or office building into existence out of this invisible material they call *the substance* that materializes upon command. You can do and make things using English as well but the big stuff the fun stuff is all done in Hebrew. Pieces that are pre-credesigned are moved into place by people called *movers*. The biggest moving company there is run by two ten year old twins and their mom. They were part of the original training group taught by the Us."

"This is amazing," Slater said, shaking her head. "Did you say ten year-olds?"

"Yeah. They were only nine when they first came to the planet. Kids are the fastest learners on the planet."

"So they were a part of Jerry Jergensen's original group?"

"Yeah…" Killix's eyes dropped from the President's gaze and he began to fiddle with his paper napkin, uncomfortable. The mention of Jerry's name always made him feel uncomfortable. Uncomfortable and guilty. "Before he got booted back here. Pretty sad, especially since he started it all."

"I think that was a mistake made basically be one man," Slater reassured him. "General Maddox!"

"Really?" said Killix. "I think it's a mistake to place something as powerful as a planet in the hands of one man. Jergensen, from my understanding, could never see the bigger picture."

"Who told you that? General Maddox?"

"No," Killix said nervously. "No, just rumors. I take it you're not too fond of the General."

"I'm not too fond of anyone who *appears* to have their own agenda when it comes to this nation. Do you know General Maddox?"

"No," the Secretary shook his head. "I only met him at my confirmation. Seemed like…a General."

Both of them laughed at this and President Slater nodded, unconcerned. "I'm watching him," she said. "No bias. Just watching."

"Really. Should I take it that you're watching me too?"

"I don't know," Slater said. "Should I be?"

She stared at him across the table for a moment, their eyes locking in an intense stare. Then she began to laugh and Killix joined her. A few seconds later the proprietor of the restaurant walked over with their steaming pizza and a smile on his face and they began their meal.

Jergensen Residence — Westerville, Ohio, U.S.A.
Earth Date: December 9th, 2015 — 1:06pm EST

As he did more and more often these days, Jerry was spending his afternoon on the couch in front of the TV. Initially after being removed from the project, he occupied his time with speaking engagements and writing. But lately, he had become bitter as he meditated more and more on the unfair fact that his discovery had been stripped away from him. As he surfed through the channels, he couldn't seem to escape coverage of Dabar. The whole country—no the whole *planet*—seemed to be talking about the parallel world that *he had found*. And yet here he was, stuck at home watching it on the television.

"*And in other news,*" one of the UNN anchors stated as Jerry paused on the network, "*it appears we may have finally discovered the fountain of youth. Only it's just not on this planet. Hundreds of visitors who have spent extended periods of time on the planet Dabar have been experiencing everything from perkier and firmer breasts to tighter and more youthful looking skin. Coincidence? Some don't believe so. We talked to a Ms. Oretha Jankowski, who was one of the original visitors to the planet and, if pictures are to be believed, she's here live with us and appearing almost twenty years younger.*"

As the news segment continued, Candy came down the stairs in her bathrobe, hair still wet from the shower she had just taken. She stopped in the doorway, leaning against the frame as she spotted Ms. Jankowski on the screen. "Oh look, Jerry! It's Ms. J! Wow!! She looks even younger than when we left!"

"I know, good for her." he said miserably. He reached out with the remote and thumbed the television off. "Sometimes I wonder, was it a good idea or not to even make the discovery?"

"Jerry, it's out of your hands now. The government has practically taken over the operations of the facility and the planet. And you know the whole fountain of youth thing is true… I mean, just look at me. Or just look at these."

She untied the belt of her robe and opened it to reveal what was beneath it. Jerry's mouth spread into a juvenile grin as he enjoyed the view. "Those are nice... *nicer*... I mean, they were always nice, they've just...gone back to the future. Or the past..."

Candy laughed at his flustered response as she closed her robe and tied the belt back around her waist.

Jerry shook his head, still feeling down. "It's just... I just don't like where this is going. It seems like things are so far out of control. The training should be getting better and more advanced, but from everything I can see it's getting sloppier. More people are dying and they're covering it up just like they did the number of military suicides. They're just pushing people through the system in an effort to make more money."

"Jerry, what do you mean?"

"Candy, that's really what this is all about. That's why they are allowing more countries to come to Dabar. Every Dimensional Mirror they sell to a country is worth millions—or even *billions*—of dollars from the initial sale and the residual revenue each country has to pay us for access."

"Are you serious?" Candy asked. "Jerry, you never told me that."

"And as far as anyone else is concerned, you don't know now. They have Calvin and I under a gag order. If we tell anyone and word gets out, they can reinstate the charges for the deaths from the original insertion and lock me and my team up for life. I'm not concerned about myself, but there are other people involved. It's all turned into a big cover up, a big lie. The United States of America on Dabar has simply become the United States of America, complete with all the corruption, greed, lies and scandals."

Then, with nothing better to do, Jerry switched the TV back on. "And it's all because of me," he mumbled. "All because I made this terrible discovery..."

Candy cut him off. "Jerry!" she said cupping his face, "This is not your fault and Jerry you don't do *pity*! They took something pure, something good, something great and perverted it. Don't you dare blame yourself." she said halfway chastising her spouse. "The truth will prevail. The Us appear weak and docile but their not!"

"But it's been a year almost and it looks like they've done nothing."

Jerry retorted, disappointed.

"Jerry, you of all people should know better! Looks are deceiving! In the end they are not going to win. You can be sure of that." Candy declared. "Don't let what they are doing cause you to unlearn what you… what we… have learned." And with that Candy left Jerry there to think and remember all that he, that they had learned.

Troyconics Astrophysics — Columbus, Ohio, U.S.A.
Earth Date: December 9th, 2015 — 3:14pm EST

Although there was only a single person in the Troyconics conference room, it had become the meeting place of several international powers. General Maddox sat alone at the head of the table amidst the newly outfitted conference room. However, on a series of screens mounted along the walls of the room, was the digital busts of several politicians. Along with Ambassadors from Australia, China, Japan, Taiwan, Brazil and South Korea, the General had also looped Secretary Killix in on the videoconference.

"Ambassadors," the General was telling the representatives of foreign nations, "our price is firm at four hundred million dollars for China and Japan, and four hundred fifty million dollars for Taiwan and Brazil for each Dimensional Mirror. And we believe the one hundred fifty million dollar discount respectively is more than generous."

"*With all due respect, General and Mr. Secretary,*" said the Chinese Ambassador from one of the monitors, "*given the fact that your country owes us over one trillion dollars, one would think that an even greater discount is in order.*"

"*Ambassador Chang,*" Secretary Killix jumped in, "*please realize the opportunity this will bring your country. We are positive that each of you respectively will make back your investment cost within one year. Now, in addition to providing you with the Dimensional Mirrors, we have also agreed to provide you with training and support for three months.*"

On another screen, the Japanese Ambassador shook his head with disdain. "*Mr. Secretary,*" he said. "*Please do not take us for fools. We realize that part of the only reason you have agreed to provide this training is to maintain oversight of what we do in our designated land on the planet as well as see to it that things are done—as you say in your country—'the American way.'*"

Calmly, Secretary Killix raised a single hand, easing the man's concerns. "*Ambassador Xui, I assure you that, within reason, you will have as much autonomy in your country on Dabar as you do here on Earth. Pending of course any hostile acts."*

The Japanese Ambassador smiled and nodded. *"Of course."*

"Countries as large as any of ours will require multiple gates," Ambassador Chang put forth. *"With our countries and populations as expansive as they are, one gate simply will not do."*

"We will require at least four," said the ambassador from Brazil on another screen.

"As will we," added the representative of Taiwan.

Ambassador Xui stared into the camera stoically. *"Our country will require anywhere from four to six."*

"Well congratulations, Mr. Secretary," Ambassador Chang announced. *"From my calculations you have just become the salesman of the planet. Most salesmen would kill for a sale worth over seven trillion dollars."*

"Gentlemen," Killix said, *"This is the Berkshire Hathaway stock of the planet. You're guaranteed a return on investment. Of course, as we mentioned earlier, you can subtract sixty percent of the debt owed to each of your individual countries. The rest is due in cash. I'll have my people send each of you the necessary paperwork."*

General Maddox smiled at the various screens. "Gentlemen," he said, "it's been a pleasure doing business with each of you."

One by one the ambassadors signed off, each screen that framed their faces turning black as each Ambassador cut the connection. Soon there were only two faces left up on the screens: the Brazilian ambassador and Richard Killix.

"We look forward to doing business together," the Brazilian ambassador said before signing off himself.

When it was only the two of them on the video conference call, the General smiled at Killix through the screen. "Congratulations Mr. Secretary! You just wiped out half of the national debt!"

Killix gave a sarcastic grin and shook his head. *"Do you believe our national debt is really only fourteen trillion? Half? You're kidding right?"*

Dabarian Justice Center — Dabar
Date & Time Unknown

The Justice Center stood in the center of the United States of America on Dabar. Made of marble that had been created from nothing but The Substance, the building was gorgeous and majestic. It stood as both a symbol of justice, and as a symbol of the beautiful things that could be built on Dabar.

Dabarian peace officers milled in and out of the building, climbing up and down the large marble steps that adorned the front of the building. Most of them had offenders with them, taking them inside for processing and detention if needed.

Two officers, Donald Tate and Charles White, were headed inside with their collar, a young man whom Tate was forced to cuff and fit with a chastity muzzle. It was the young man that had given the ThreeofUs the finger. The man who the ThreeofUs had called 'Katapugon.'

Together, the two partners took him inside and pulled the Katapugon over to the processing desk where an older man, Officer Mason Grey waited. "And what do we have on the menu tonight, gentlemen?" Grey wanted to know, eyeing the cuffed and muzzled man.

"Well, Mason, this lovely gentleman was caught on infrared camera in a movie theater removing his pants and sticking his middle finger up…up his butt…repeatedly."

"Yeah, he was asked to stop and leave, but refused. And so here we are."

"Great!" Grey said, turning to the Katapugon. "What's your name, Son?"

"Oh," Officer White said, "we forgot. He's not talking, but he did do us a favor and left us a note to tell us who or what he is."

White reached out and unbuttoned the young man's button-down shirt, pulling apart the sides to show Grey what lay beneath. There, written on the young man's chest in permanent marker, and with excellent penmanship, were four words: "I AM A KATAPUGON."

"Okay, Mr. Katapugon," Officer Grey said. "Let's get your prints and then we'll find out who you really are."

"Hey Mason," Tate said, "what is it? Look it up."

"What?" Grey asked, irritated at being bothered further.

"Katapu...whatever. What is it?"

Sighing, Grey turned back to his computer and typed in the word, eyes flicking back to the young man's chest to check his spelling. "Okay!" he finally said. "This explains it."

"What?" asked Officer White.

"A katapugon is the name of the gesture people make with the middle finger."

"You mean flipping the bird?" Tate asked.

"Yeah, but there's more. It also means 'a male who submits to anal penetration.'"

White couldn't help but laugh. "Well, Mr. Katapugon, you've come to the right place. You'll love jail once we ship you back to earth."

The officers shared in the laughter. The Katapugon however, was crying. His tears ran down his face, only to seep under his chastity muzzle and onto his lips so that he could taste the salt of them. He tried to speak, but found that he could only make muffled sounds through the muzzle.

"Take it off," Grey said. "Let him speak. How long have you been in training son?"

Tate obliged "I only have a week left," the Katapugon said, "I know what to say and not say. He did this," the man insisted. "I didn't do this."

"Oh, so you do talk," White said. "Don't say anything crazy! But c'mon man, they got you on video."

"I know, but I didn't do this. The man. The man behind the gate..."

"Yeah buddy, sure," said Officer Grey. "Other people are forcing your fingers up your butt. Right thumb please." He reached out and wrapped his fingers around the katapugon's right wrist, pulling it close to get his print.

"Hey Mason," Officer White laughed, "you better watch out. He's right handed!"

Grey let go of the Katapugon instantly. "That's disgusting!" he cried as he reached for an antibacterial wipe. The officers continued to laugh as they continued to process the Katapugon.

The Sanctuary of Us — Dabar
Date & Time Unknown

Deep in the bowels of the Dabarian castle, in a room that Jerry or any of the other trainees had never seen, red light spilled from the seams of a tubular door. The door was one of twelve, and as each tube emitted its own red light, it filled the room with a crimson gloom reminiscent of a photographer's dark room. Soon the curved doors began to slide open, letting more light out as the Us stepped out of each tube.

The ThreeofUs left his tube and stood in the middle of the room as the other eleven of Us surrounded him, forming a circle with him in the center.

"What is it?" the other eleven of Us asked him in unison, their individual voices becoming one in the rose colored chamber.

"A group of visitors today entered the gate briefly," the ThreeofUs explained. "We stared at them and they exited the property."

"But?" the others asked.

"One of them attempted to curse Us with a gesture."

"We cannot be cursed," the Us pointed out.

"We said so be it," the ThreeofUs said, "and called him verbally what the gesture meant."

The Us closed their eyes briefly, their minds simultaneously twelve and one. After a few moments, their eyes opened again and all of them spoke together: "So be it," they said. "He is katapugon."

Dabarian Justice Center — Dabar
Date & Time Unknown

"All rise!" called the bailiff and everyone in the courtroom got to their feet. The large room was crowded, each row of benches packed with citizens of Dabar. Sitting in the defendant's chair, the Katapugon wore a blue jumpsuit and chastity muzzle. Unable to speak, he looked back at the audience. Amongst them, his father sat, a disapproving look plastered on the man's face. "The honorable Judge Amy Twiss presiding!"

The Judge entered the courtroom, her black robes flowing. "Be seated," she instructed the court after she had settled behind the bench, and the audience sat back in their chairs.

The Bailiff stepped forward, unlocking the Katapugon's chastity muzzle and removing it. "First case," he called out, "is the people versus Mark Grammar with two counts of lewd and indecent conduct."

Judge Twiss nodded to the Prosecutor. "You may present your case."

"Your Honor," the Prosecutor said, getting out of his seat and pacing about the courtroom floor, "the Defendant was caught on infrared video in a movie theater with his pants down and repeatedly inserting his finger into his rectum."

As the prosecuting attorney laid out the case, the Katapugon's father looked away from his son, unable to maintain eye contact. Ashamed, the Katapugon looked down at his feet.

"The video speaks for itself," the Prosecutor added, gesturing to a screen that had been built into the sidewall of the courtroom. On it, the infrared surveillance video began to play. Disgusted murmurs and immature laughter rippled through the courtroom and the Katapugon sank down in his seat, embarrassed.

The Judge looked away from the screen, waving the image away. "I've seen enough. Defense?"

The Katapugon's defense attorney stood up. "Your Honor, we have nothing to add. The Defendant was on prescription drugs that reacted unfavorably with antibiotics that he was on which resulted in a hallucinogenic response from my client. We plead guilty." The bailiff then walked over and removed the chastity muzzle to allow the Katapugon to plead.

"Well, I am happy to see that you and your client have not decided to waste any of the court's time," the Judge said. "Mr. Grammar, I see you have never had a run-in with the law, so I am inclined to believe your attorney. That this was an isolated incident. The Defendant is guilty as charged and is sentenced to six months probation and community service at a local church. Any objections?"

"None, Your Honor," the Prosecutor said.

The Katapugon's attorney shook his head as well. "None, Your Honor."

And is this agreeable to you, Mr. Grammar?"

The Katapugon nodded. "Sure. Yes Ma'am."

"Good. I don't want to see you back here, Mr. Grammar. Otherwise there will be jail time. Are we clear?"

"Yes Ma'am," the Katapugon said.

"Great! Have a good day then," the Judge said and then slammed her gavel with deafening finality. The Bailiff ushered the Katapugon out of the courtroom, handing him off to more Peace Officers, who would take him through the release process. Then, as the bailiff called the next case to court, the Katapugon's father rose from his seat and walked out of the courtroom.

Outside, he waited in the lobby another fifteen minutes until finally his son returned. No longer garbed in the institutional blue jumpsuit, the young man was back in his own clothes. "Dad," he said with desperate relief. "Thanks for coming."

He stepped forward to hug his father, but instead the older man slapped him across the face. The other people in the lobby turned and stared as the Katapugon pressed a hand against his stinging cheek. His eyes were brimming with surprise and hurt.

"What were you thinking?" his father hissed. What were you doing? Are you gay? Please tell me you're not gay."

"I'm not gay," the Katapugon said. "It wasn't my fault—"

"Really? Then whose fault was it?" his father asked. The Katapugon thought back to the incident at the gate. He knew who had done this to him.

The Sanctuary of Us — Dabar
Date & Time Unknown

Watching from his usual spot on the castle's terrace, the ThreeofUs watched the now-ordinary hustle and bustle that swarmed at a distance from the gate below. He watched as a young man—the same young man that had tried to curse the ThreeofUs—made his way up to the gate. The ThreeofUs noticed his muzzle was removed.

"What did you do to me?" the Katapugon yelled up to the ThreeofUs from the ground, craning his neck to look at the being atop the balcony. "I know it was you! What did you do to me?"

"I would be watchful using your mouth in your currently elevated emotional state," the ThreeofUs warned him. "The outcome could prove to be more undesirable than your initial gesture."

The Katapugon closed his mouth for a moment, gathering his thoughts and focusing on the words he was using. "What did you do to me?" he repeated, more calmly this time.

"*We did nothing to you,*" the ThreeofUs proclaimed. "The correct question is what did you do to yourself?"

"I didn't do anything," the young man insisted. "I didn't say anything to you."

"Mr. Grammar, communication in most cases is almost always more non-verbal than verbal. You believe your lack of verbal communication can deceive everyone into believing you have done nothing. You have been, as many of your people have been, misinformed. Your gesture was noticed and deflected back to you."

"What do you mean 'deflected?'" the Katapugon demanded.

"We cannot be cursed," ThreeofUs explained. "Anyone who tries to curse Us in essence curses themselves. As the officer explained to you earlier, the gesture you made toward Us has a name. It is called *katapugon.*"

"How do you know that? How did you know about the Officer?"

"That finger gesture was a curse. An insult that, when directed toward another person, is the equivalent of saying or calling that person katapugon, which means 'a male who submits to anal penetration.' Mr. Grammar, you should be thankful that it was simply *you* who penetrated yourself."

"You did it to me!" the Katapugon yelled. "I had no idea what a catapult or whatever you call it was before today."

"Really? Your gesture was directed to us and came back on you. Your ignorance of what you were saying or doing is of no consequence here or on your planet. The effects of your actions on Earth are simply not immediate and therefore your people continue in their foolishness. All language, verbal or non-verbal, has effect in both places, be it positive or negative. Regardless, Our words were not without purpose. If not for the incident you would not be here now."

"What purpose?" asked the Katapugon, confused.

"You are here to become responsible," the ThreeofUs told him.

"Responsible for what?"

The ThreeofUs did not answer. Instead, he disappeared beyond the darkened archway of the Sanctuary. The Katapugon looked up at the empty balcony, his emotions a churning mixture of anger, confusion and humility. "Where are you?" he called out, demanding answers to his questions.

"We are right here," the ThreeofUs voice said from behind him.

The Katapugon whirled around to find the ThreeofUs standing right behind him on the ground. "How'd you do that?" the Katapugon asked.

"I did nothing," the ThreeofUs said. "*We* did. However, the question is not how We did this or that. The question is: *why can you not?* As we stated earlier, you are here to become responsible."

"For…"

The ThreeofUs moved closer to him, bringing his face only inches from the young man's. "Every person has a given responsibility to *be dangerous*…by being who they really are! Not who they think they are, but who they know they are."

"Dangerous to what?"

"Dangerous to what? Dangerous to average. Dangerous to mediocrity. Dangerous to normal. Dangerous to evil. All these things I have mentioned *are* evil."

The Katapugon stared back at the ThreeofUs in silence. He was just beginning to understand. "Teach me." he said in earnest

Jiasin HQ Creative Site — Dabar
Date & Time Unkown

Standing before a large empty plot of Dabarian land, Changyi Yelmi made her way to each member of her cre-struction team, passing out tablets with their individual assignments preloaded on them. Changyi was the Chief Cresign Project Manager for the Kenyali Creative Group, and it was her job to turn the new Jiasan Dabar Headquarters from an idea into a reality.

She mentally called role as she walked down the line, distributing the tablets. There was Kelma Sayerville, their Interior Specialist and her twin brother Keltay, the Exterior Specialist; there was Brad Meier, the Structural Engineer and Efrat Shabat, the seventy-three year old Architect and Hebrew Specialist who, after spending a few months on Dabar, now looked to be closer to forty-five years of age.

Finally, Changyi came to Max and Mack, the team's movers. At the moment they were busy clearing out the foundation of the empty lot. Mumbling under their breath, the two brothers stared at the cre-struction site, lifting a large bulk of dirt, rock and foliage with the power of their words and depositing it several hundred yards away.

"They let you guys out of school early so you could go to work?" Laughed Brad, cracking on the young boys.

"Keep laughing," Mack told him jovially. "I'm only ten and I've already made ten million dollars. How much have you made?"

"You got a pretty smart mouth for a ten year-old," Brad pointed out with a grimace.

"A smart mouth?" Max asked. "Or a true mouth?"

"And now there's two of you," Brad said. "Great."

"Brad, get over yourself," Changyi told him, handing Max and Mack their respective tablets. "There's a reason they're the number one movers on the planet." Then she turned to the boys. "You guys ready?"

"Yes Ma'am," they said together.

"See?" Changyi said to Brad. "You get respect when you give it."

Brad rolled his eyes and opened his mouth to say something, but his voice was squashed by the blaring sound of a bullhorn. "*Alright,*" the Foreman bellowed. "*We need everyone to clear the creative field! Thank you.*"

"Okay, let's do roll call. Kelma, you have interior?"

"Check."

"Keltay, you've got exterior?"

"Check."

"Brad, you're in charge of structural and cresign integrity."

"That's me," Brad replied.

"And Efrat, you're our new architect and audible on this project," Changyi welcomed him to the team. "I hope you're as good as they say you are."

"I am," Efrat said. "And I am."

"And Max and Mack, you are our movers and shakers... Well, not shakers, but you know what I mean."

"We're good," they said in unison.

"Good," Changyi said. "Max and Mack have already cleared the foundation. Everybody, let's do one final sync check and measurement match, and after that we'll get started."

The team came together, standing in a circle with their tablets out. Each of them tapped a few commands to data-sync their interfaces. As they linked up the tablets, each screen turned a bright green—the ready status— then switched over to red.

"Okay everyone," Changyi instructed, "we'll do this by the numbers. When your tablet screen is red that means it's not your turn yet. When your tablet screen turns green that means you're up. Approximate creation time is seventeen minutes." Then she spoke into her comm link: "Foreman, are we clear?"

"*We are clear,*" the foreman informed them all.

"Great. Keltay, you're up."

Keltay's tablet turned green and he took a step forward. Shutting his eyes and focusing every thought on the task at hand, he began to whisper under his breath. "Yehi musad—foundation; im mesequeth—pipes; im azar—girding; im gadar kachal—build walls painted; tsaphith—cloth or carpet; arubbah—windows; im qarqas, im shequeph—frames; im deleth—doors."

As he spoke, his voice rising as he spoke the Hebrew words, a few specks of what seemed to be black sand appeared in the air above the foundation. Swirling in a tornado-like vortex, the sand formed a spinning cloud of The Substance. Soon the cloud seemed to descend, covering the entire dugout of the creative site with a concrete foundation, complete with pipes, drains and outages. As he continued to murmur the words, eyes closed, the building began to take shape in the form of steel girders and supports. Steel pan with concrete floors appeared. Painted walls and carpeted floors rose replacing the black swirl.

A few feet away from Keltay, his sister's tablet lit up green and she joined her brother in his verbal symphony. "Yehi appiryon…"

They spoke together in tandem, forming both the inside and the outside of the building simultaneously as they continued to speak Hebrew commands. Finally the outer structure of the building was finished and the twins' tablets went red again.

"Good," Changyi said. "Brad, can you give me a structural integrity check?"

Using his tablet, Brad activated the structural laser scanner, which swept over the brand new building's skeleton with red light. The automated lasers, which were placed at each corner of the foundation and looked like surface-to-air missile launchers, penetrated the foundation and inner structure.

When they were finished, Brad nodded. "It's real and it's sound."

Keltay and Kelma's tablets went green again and they continued uttering the Hebrew words. As they did so, Efrat watched on his screen, making sure that they weren't missing or adding any words that didn't belong. Soon the building was taking its full shape and color and the twins began adding more minute details.

"Okay Max and Mack, you're up."

Ready for work, Max and Mack had already started moving the large thermo-conditioning units and solar panels that would be installed on the roof that had already been created. "Nasaq-(Ascend)," Max said.

Together the two brothers lifted the units like living cranes. While Max controlled how high they went, Mack moved them forward and back, and to the left or right. "Ramas," he murmured. "Creep or move; yaman—(to the right)."

The units soared high above them, sliding into their proper coordinates as Max and Mack expertly maneuvered them into place.

"Okay Mack," Changyi called out, "not too close to the building. We don't want the units in the building, we want them *on* the building."

"We got this," said Mack.

"Nasaq—(ascend)," Max said, "yachid—(only)!"

The units stopped sliding to the right and shot straight up, headed for the roof. As the rest of the team watched the units soar high above them in the air, a small blur quickly passed by the units. It was Mack. Slowing himself and maintaining a height slightly above the roof of the building he slowly continued positioning the units.

"Ramas—(move or creep)," he said. "Yaman—(right)."

He deftly settled each unit into their final resting place on the roof. When they were exactly where they were supposed to be, Mack lowered his arms gently and stopped the units with a single word: "yashab-(remain)."

When he was finished, Mack still standing only on air examined the placement closer. Down below, his brother and the rest of the team watched him intently as he turned his palms down and whispered, "Ramas—(move or creep); nacheth—(go down or descend)."

He slowly began to lower himself down to the ground, examining the outside and inside of the building as he passed each new floor. "We're good here!" Mack told Changyi over the comm link as he gently reached the ground.

When he walked over to the rest of the team, they were looking at him with wide eyes. "Wow!" Changyi exclaimed. "I've never seen you do that before."

Mack smiled. "Yeah, I started thinking. If I can move other things a lot heavier than me, I can probably move myself."

"Yeah…" Changyi said, still shocked. "Right. Okay, so Brad and Efrat, can you give me a complete process and structure check?"

"We're on it," Brad told her.

"Okay Kelma. Let's go see how good you did." They walked towards the building entrance.

Kelma flashed Changyi a grin. "If there's on chair missing I'll buy you a drink!"

"So confident," Changyi smiled back. "I love it. Makes my job so much easier."

The rest of the team began their structural integrity post test. Many miles away the OneofUs stood on the top of the castle. The expressionless face seemed to have the faintest hint of a smile. "Our dear Mack and Max." he said as if he were a proud father and quickly turned and stepped into the darkness of the archway.

Troyconics Astrophysics — Columbus, Ohio, U.S.A.
Earth Date: December 12th, 2015 — 8:26pm EST

Back in the conference room, General Maddox sat with Calvin and Bernard Maychoff as they video-conferenced with President Slater and the Joint Chiefs on the multiple screens.

"*Ladies and gentlemen,*" the President addressed them, "*I want us to seriously consider what we really need and what is simply greed. General Maddox and Mr. Killix, I believe congratulations are in order for your most recent work on securing the Dimensional Mirror sales to Russia, China, Brazil and Taiwan. With those sales of approximately seven trillion dollars and the others, the United States will soon be a debt free country with a surplus for the first time in years.*"

"*Madam President,*" Secretary Killix said, "*if I may, don't allow that influx of cash to get you too comfortable. With rising cost and our need to replace some of the defense cuts we recently implemented, we could very easily find ourselves back in debt very quickly without additional revenue sources.*"

"With all due respect," Calvin interjected, "this discovery was not made to be an income source. It was made for the purpose of exploration and discovery."

"*Calvin, I respect your position for science and all,*" Killix replied, "*but in light of the fact that we have already opened up the planet to other countries, there are distinct possibilities that although we have made the initial discovery, we could be quickly surpassed by others if we do not capitalize on it.*"

"*And what exactly does 'capitalize on it' mean?*" the President asked.

"*Madam President, I am proposing we acquire additional acreage. With more land we can then lease that land out to the other countries.*"

"And exactly how do you plan to get the Us to agree to that?" Calvin challenged.

"With all due respect," Maddox said with a grin, "we've been very conciliatory with the Us."

"As we should!" Calvin shot back. "It's their planet!"

But Maddox seemed to ignore Calvin's comment altogether. "With our current rate of occupation," he said, "we'll be out of space within a month with no room for expansion. We will require an expansion at least five times our current size for our own people and then additional space for acquisitions in the future."

"*We didn't anticipate such rapid growth,*" Killix added, "*especially before releasing Mr. Jergensen from his position. We have not had contact with the Us since he and his wife left the planet.*"

"*In the event that we were to forcibly take additional land from them,*" posited the Secretary of Defense, "*what would be their response? What are their capabilities?*"

"From everything we've seen," answered General Maddox, "minimal at best. We've even pushed a little bit and all they do is talk."

"*From everything I understand, General,*" President Slater said, "*all they* have *to do is talk.*"

General Maddox shook his head. "They don't believe in violence or negative behavior in any form, so I think it's safe to say physical attack or repercussions are not a concern."

"*My recommendation is that we try to use diplomacy,*" said the Defense Secretary. "*But be prepared to use force if necessary, if we encounter any resistance.*"

"*Agreed,*" said Killix.

"*Let's not be to hasty,*" President Slater said. "*I don't like the idea of forcibly taking over someone else's planet.*"

"Madam President," argued General Maddox, "this is exactly why I had Jergensen removed from the planet. No one has taken the time to acknowledge that *it's not their planet!* They discovered it just like us. They *were* people just like us."

"*With all due respect General,*" the President told him, "*they made the discovery first. They were there first!*"

"And that gives them the right to dictate what can and can't be done with an entire planet?"

"*They were there first!*" Slater said, her voice suddenly taking on the weight and strength of the Presidency. "*I don't know how this doesn't reek of the Native Americans and Pilgrims to you all over again."*

"Yeah," said the General, "and if we had left the Indians in charge we'd still be living in tee-pees and smoking bad weed. *With all due respect, Ma'am."*

"As opposed to the good weed you've obviously been smoking?" Slater asked, letting the wit of the statement wash over the rest of them. *"General, your waspishness is showing."*

"And what's a wasp without a sting?" Maddox spat back quickly. The room tightened and became quiet.

"Is that contempt I see on your face, General?" Slater demanded.

"No Ma'am," Maddox swallowed his words. "I'm simply facing the facts."

As the President and General Maddox continued their verbal jabbing and parrying, one of the Joint Chief's assistants leaned over and whispered in his ear. "Sir," he asked, "what's a wasp?"

"I don't know what it means today," his boss answered, "but it used to mean white Anglo-Saxon Protestant."

"Oh!" the assistant whispered, nodding. Then after pausing for a few seconds, "What's a white Anglo-Saxon—"?

"Son, Google it!" the Chief yelled before returning his attention to the videoconference.

"We'll reconvene tomorrow morning," President Slater ordered from her screen. *"I need to sleep on this. I suggest some of you do the same."* As she spoke, she eyeballed the General with a raised eyebrow. He stared back intently, refusing to give up any ground. *"Everyone have a good night."*

Her screen went dark and soon, one by one, the Joint Chiefs signed off as well. Before they knew it, General Maddox, Calvin, and Maychoff were left alone in the empty room with the darkened screens. "See?" the General growled. "This is what happens when you put a woman in office."

Calvin shook his head. "You're wrong, General. But unfortunately you won't see it until it's too late." He stood up from his chair and left the conference room. Maddox left soon after and then only Maychoff remained in the conference room. He sat there, leaning back in his seat and stroking his chin, deep in thought.

The Sanctuary of Us — Dabar
Date & Time Unknown

While Max, Mack and the rest of Changyi's cre-struction crew put the finishing touches on the new Jiasin Dabar headquarters, Mark Grammar—the Katapugon—continued his conversation with the ThreeofUs. They looked out at the ever growing skyline of the United States of America on Dabar.

"It is evident that there is still a great affinity for tall structures amongst your people," the ThreeofUs said as Mark joined him on the terrace.

Mark simply shook his head in wonder. "I don't understand how they build them so fast."

"Would you like to find out?" the ThreeofUs offered. "Would you like to discover our purpose? Your purpose?"

Mark turned to the ThreeofUs and stared at the strange being's teenage face, but he said nothing.

"Join Us," the ThreeofUs went on, "and all will be made known to Us."

"What do I have to do?" Mark asked.

"Believe!" the ThreeofUs proclaimed. "On your world they have a saying. Although we have not been there for many years, we are positive it is still prevalent. They say 'seeing is believing.' However, here we prefer to go the other way, for we know that 'believing is seeing!'"

"How can you do that?" asked Mark. "How can you believe in something you cannot see?"

"You do it every day. Do you believe you have a brain?"

"...yeah?"

"Have you ever seen it?"

Mark paused, realizing what the ThreeofUs was getting at.

"Of course not," the ThreeofUs answered for him.

"I see your point."

"Good." the ThreeofUs said. "Shall we begin?" Then the ThreeofUs walked away from Mark and disappeared into the Sanctuary. Mark looked around, unsure of what to do at first. Then he shrugged and followed the being inside. He had somewhat unknowingly but officially become one of Us.

The White House — Washington D.C., U.S.A.
Earth Date: December 13th, 2015 — 2:33pm EST

As the Joint Chiefs of Staff and a host of other military personnel settled into their seats in the White House briefing room, along with a handful of NSA agents, General Maddox brought the meeting to order.

"Gentlemen and ladies," he began. "I believe you've all been briefed, but I will give you a recap. We are here today to provide the President with a detailed plan, which includes strategy of attack and threat level assessment in an effort to obtain authorization to invade and use lethal force if necessary on the planet Dabar. Mr. Killix?"

Killix nodded and took over. "We realize this scenario is unfamiliar to you, however based on the information you have each been provided with, we are asking for detailed information from each of you so that we may create a proposal for an executive order that the President can approve of for us to act on."

"This is tantamount to declaring war on a nation," one of the Joint Chiefs said, his voice thick with reservations and concerns.

"In a way it is," the General responded, "and in a way it isn't. This is not a nation. This is a group of roughly now a few hundred individuals whose core group has assigned themselves as the overseers of an entire planet. So I don't think that qualifies as a nation. We have a considerably aggressive timetable so in light of that we have assigned other capable individuals to oversee your daily duties. Make no mistake this is priority number one. We need everyone's report in three days so that we can condense the information for our proposal to the President.

"In the meantime," he went on, "Secretary Killix and I will be visiting Dabar tomorrow to request a date for a meeting where we can discuss a diplomatic means to acquire more land. The President has been informed and we need to have this land agreement in place before or by the December deadline in order to meet our agreements and conditions listed as part of the sale of additional mirrors to Brazil, Japan, China and Taiwan. Any questions?"

"What if things do not go as planned?" Another Joint Chief asked.

"That's why we're here today. So that we have an immediate contingency plan ready in place."

"What is the Us threat level?" asked the Defense Secretary.

"From our perspective, it is a one or two," Maddox told him. "They are completely and adamantly against physical violence of any sort, so based on that we don't anticipate much—if anything—in the form of resistance as we know it."

"So you haven't confirmed that?"

"Not as of yet, Admiral," interjected Secretary Killix. "Knowledge and information on the Us is sketchy at best. However, we are in the process of meeting with someone who has the closest knowledge of them, at least on this planet."

Troyconics Astrophysics — Columbus, Ohio, U.S.A.
Earth Date: December 13th, 2015 — 4:56pm EST

Calvin sat behind his desk, asleep in his chair. Soft snores exited his open mouth with each slumbering breath and his head hung to the side at an uncomfortable angle. Suddenly, the phone began to blare and Calvin woke from his unplanned nap with a start. Blinking the sleep from his eyes, Calvin got his bearings then answered the phone. "Calvin Madison."

"*Mr. Madison,*" the General's voice came through the line. "*Did I wake you? I thought you were at work.*"

"I am," Calvin said. "How can I help you, General?"

"*Mr. Killix and I need to visit Dabar tomorrow afternoon. We'll be going in through the original insertion point.*"

"Okay."

"*If you could have everything prepped and ready for us to do so by 1300 hours—*"

"Not a problem, General," Calvin said. "Is there anything else?"

"*As a matter of fact there is. We need to meet with Jerry Jergensen.*"

"For what?" Calvin asked. "You took away his project. You kicked him off of Dabar. What do you want now?"

"*We just need to ask him some questions. It's a matter of national security. Can you arrange to have him meet you in the conference room? We'll show up a few minutes later.*"

"So lie to him," Calvin sighed.

"*No,*" said the General, "*just tell him he will be meeting with you. He does not have to know about the other people.*"

"This is the last time, General. Find someone else to do your dirty work."

Madison, just make sure he's there by eleven hundred or you can look for a job too. And at your age, you'll need a miracle if we decide to blacklist you. We'll see you tomorrow!

Then there was a harsh click on the other end of the phone and the line went dead. Calvin held the receiver away from his ear, staring at the earpiece. Then he slammed the phone down in its cradle.

Chapter 11

Five black Chevy Silverados cruised down the highway in perfect formation. Their windows were tinted and their license plates were simple government decals. In the middle car, General Maddox and Secretary Killix sat in the back, continuing their contingency plan as the convoy carried them from Washington D.C. back to Ohio and the Troyconics facilities.

"If I could find someone else I would replace him too," the General was lamenting of Calvin. "He and Jergensen are a pain in the ass—"

"—set!" Killix finished. "Jergensen and Madison have both been assets in this game of checkers. Without them both, none of this happens."

"No one asked you for the bright side," the General grumbled.

"So is the meeting set?"

Maddox nodded. "He'll be there. Once we talk to Jergensen we'll both go in to Dabar and request a meeting with the Us for December 20th. That way, regardless of what happens, it will give us time to run the executive order by the President and set up the execution of our contingency plan."

Secretary Killix began to laugh to himself, staring out the window as the farmlands of West Virginia flew past.

"What's funny?" Maddox asked.

"Nothing," the Secretary said, shaking his head. "It's just the appearance of compliance is almost worse than the non-compliance."

The General didn't share in Killix's amusement. He leaned forward, looking the man in the eyes. "We're saving the country here. I served my time in Vietnam and other wars and there's a lot of truth to the saying: 'in order to make an omelet, you gotta break a few eggs.' Well the Us are the eggs."

"I'm on your side," Killix reminded him. "I don't have to like it, General, but I see your point."

"Good," the General nodded and sat back in his seat as the car brought them closer to their destination.

Jergensen Residence — Westerville, Ohio, U.S.A.
Earth Date: December 14th, 2015 — 1:36am EST

"So when do I get to meet this Josh?" Kendra asked.

Alexandria sat in her room, talking to her best friend on the phone as she painted her toenails. Lexi smiled at the thought of Josh. "You've got to meet him," she said. "He is so cute."

"When does he come back from the planet?"

"He's scheduled to come back off on leave in three days," she told Kendra.

"So what does he do?" her best friend prodded.

Lexi's face lit up with pride. "He's an architect."

"Girl, shut up!" Kendra squealed with excitement and just a sliver of jealousy. *"You going for the big money! I didn't know you was a gold-digger. But you do be pimpin' yo' daddy so?"*

"No, it's not like that," Lexi laughed. "He helps create a lot of the stuff on the planet."

Kendra opened her mouth to ask more questions, but before she could there was a knock at the door and Lexi's dad popped his head in. "Lexi," Jerry said, his eyes half shut, "it's almost two in the morning. Don't you have an interview tomorrow? You should really get some rest."

"Okay Dad, I just had to give myself a pedi" she told him.

He walked over and planted a tender kiss on her forehead before leaving the room. When he was gone, Lexi put the phone back to her ear. "Kendra, I need to go."

"Is that your dad?" she asked through the phone. *"What is he still doing up?"*

"I don't know why he's still up," Lexi said, shrugging even though she knew Kendra couldn't see the gesture. "I'll talk to you tomorrow."

"Okay, Girl," Kendra said. "Later."

Meanwhile, on the other side of Lexi's closed door, Jerry was making his way down the staircase to the kitchen for a midnight snack. As he crossed through the living room, something in the dark shadows caught his eye and he flicked on the lights.

"My goodness!" he exclaimed, jumping backward in surprise. "ThreeofUs! How? What are you doing here?"

The ThreeofUs, looking strange and out of place on Earth, stepped forward out of the shadows, his teenage face looking up at Jerry with its usual blank expression of wisdom and serenity. "Jerry Jergensen," the being said, "it is once again good for Us to see you."

"But how?" Jerry asked, his heart pounding in his chest.

"Jerry, your people must learn to listen to what is not said as much as you listen to what is said. We simply told you We had not been to your planet in a long while. We did not say We could not come."

"What's going on?" Jerry wanted to know. "Why are you here? What's happened?"

"It is not so much what has happened as is what is *about* to happen."

Just then, Jerry heard the soft footsteps of Candy coming down the stairs. She rounded the corner, still half asleep. "Jerry? I heard talking. I thought—" She let out a quick gasp as she saw the ThreeofUs standing there in her living room. "ThreeofUs! How?"

"You have both been married to each other very long. You duplicate each other's speech. Good morning Candy Jergensen."

Candy took a seat on the sofa before her knees gave out from shock. "Jerry? What's going on?"

"I don't know yet," he told her.

"We do not have much time," warned the ThreeofUs. We require a favor of you Jerry. Tomorrow, General Maddox will ask to meet with you and Calvin along with a man named Richard Killix."

"For what?" Jerry asked.

"You will find out. However, they will ask if you believe if threatened that the Us will retaliate with physical violence. You will tell him 'no'. You will assure him that no, We will not. That is all We can tell you. Can you do this for Us, Jerry Jergensen?"

Jerry thought about what was being asked of him for a brief moment then nodded his head. "Sure, but why? Are they threatening Us?"

"They have not," the ThreeofUs said stepping closer to Jerry. "Yet. Do not worry. Everything is alright!" Then he reached out and touched Jerry's forehead and briefly a small glint of light appeared between his finger and Jerry's skin. "Hey! What did Us do?" Jerry said, unsure of what just happened. "You do not need to know now but,… when everyone else forgets, when the time comes, you will remember." The ThreeofUs said almost sadly but vaguely.

Then the ThreeofUs began to fade into nothing, as if he were some kind of ghost or phantasm. Jerry and Candy watched, wide-eyed as the teenage body slowly dimmed. "It is good to see you again," he told them before disappearing completely. "We will miss you."

And then he was gone, leaving Jerry and Candy all alone in the dimly lit living room. Jerry stared at the spot where the ThreeofUs had stood, his eyes drooping with sadness.

"What is it, Jerry?" Candy asked. "What's the matter?"

"They did not say We will see you soon. They did not say We will see you again.'"

Candy stood up and went to her husband, again cupping his face in her warm hands. "They also said do not worry," she reminded him. "Do you trust Them?"

"Yeah…" Jerry told her.

"Then let's go back to bed," she said, and taking him by the hand, she led him back upstairs to their room, turning out the lights as they went.

Xavier Arena — Warren, Ohio, U.S.A.
Earth Date: December 14th, 2015 — 9:16am EST

The audience clapped heartily as Jerry finished his lecture on advanced quantum physics. He had picked up a few more speaking engagements. But it wasn't the same as being in the lab. As being back on Dabar.

As he walked off the stage, he felt the buzz of his cell phone in his pocket. He dug it out and checked the screen: Calvin.

"Calvin," he answered the phone. "Long time no hear from."

"Oh Jerry, stop it. I just talked to you two days ago."

"Grumpy."

"Look, can we meet this morning here at the facility? I just need to have a few questions answered."

"Sure," Jerry said, remembering what the ThreeofUs had told him earlier that morning. "Is everything alright?"

"Everything is fine. Just need your help and it will be better if we meet in person."

"Is Maddox involved?" he asked.

"Jerry," Calvin said with a sigh. *"Just be there by ten-forty-five, okay?"*

"Sure, Calvin."

"Thanks!" Calvin said and hung up the phone.

Jerry stared at the cell phone in his hand and took a deep breath. "I guess this is it," he said.

The White House — Washington D.C., U.S.A.
Earth Date: December 14th, 2015 — 6:51pm EST

President Slater stood in the Oval Office, staring through the large windows that looked out on the lawn of the West Wing. Deep in thought, arms folded in front of her, she looked out on the capital of the nation she had vowed to lead.

Suddenly, a knock at the door brought her out of her reverie. At first she didn't answer, but when the knock came again she turned around to face the door. "Come in."

The door opened and in walked Secretary of State Hicks, holding a file folder. "Madam President," he said. "The executive order is here."

He slid the file to her across the desk as she stepped behind it. She picked it up and flipped through the papers, reading every word. "Authorization to invade and use lethal force…" she read aloud. "Hicks, how do you feel about this? What if another country were to challenge our sovereignty?"

"Madam President, this is beyond feelings. For many people this is about prosperity and simply survival. The past few years America's prosperity has been severely challenged and questioned not just by us domestically, but by people all over the world. They think we're all washed up. A has been. Many of the people involved with this decision see this as the way out of it and they don't believe such an opportunity will knock twice. If we don't take advantage of it now and we've already allowed other countries access to the planet they will. Without a doubt, other countries are developing their own plans as to how they can make advances and take advantage of Dabar and make it work for the good of their nation. They'd be fools not to."

Before Slater could respond, there was another knock at the door. "Come in," she called out.

This time it was General Maddox and Secretary Killix that entered. "Mr. Hicks," the General said by way of greeting. "Madam President."

"Madam President," Killix also nodded. "Mr. Secretary."

Slater stared the two down, allowing her body language speak for her. "Can you give me the short version?" she demanded.

"Yes Ma'am," Maddox told her. "Upon my last visit to Dabar, I visited the sanctuary along with Secretary Killix and requested that the Us meet with a representative from the United States to discuss additional land needs and revisions to our previously agreed upon terms."

"None of which are in writing?" she said.

"No Ma'am," said Maddox. "They are very strict and once they give their word they believe that is it. Unlike us no contracts are needed or executed."

Slater nodded, thinking. "If only things were that simple here. Continue."

"If I may, General?" Secretary Killix cut in.

"Sure."

"The…ThreeofUs, as he is called, did not appear surprised at all by our request and actually stated that he or they had anticipated our request for more space due to the rapid expansion and number of people who have moved to Dabar. He believed that a meeting was in order to discuss changes and adjustments, but stated that the Us conglomerate had already deemed it not in the best interest of all parties involved to allow more room at this time for expansion."

"Did he say why?" Slater demanded.

"They felt that the increased number of involuntary suicides displayed a sloppiness in training on our part," said General Maddox with a dismissive air. "And that things were being pushed too fast in an effort simply to push people through the system and make more money. If I didn't know better, I'd think they had been talking to Jergensen again."

"I wasn't aware that the number of deaths by involuntary suicide had increased," the President mused curiously. "My reports did not reflect that."

She stared hard at Secretary Killix, waiting for an explanation. "The increase has only been minimal, I assure you," he said. "The bottom line is that they are willing to meet to discuss, but for the most part they have already decided against allowing any expansion at this time."

"So did they agree to our meeting today?"

Nobody answered her. Maddox and Killix instead shared a knowing glance that President Slater couldn't help but pick up on. "What?" she asked.

"This is where it got interesting," said Secretary Killix. "When we informed the ThreeofUs of the date we would like to meet we were informed that that particular date was unavailable due to *The Rest*."

"The what?" Slater repeated. "What is that?"

"Apparently each year," the General explained, "the Us must take one day where they all go into a sleep-like state and rest *themselves*."

"They told you this?"

"They are of the mind set that they have nothing to hide," the General told her.

Sitting down behind the Presidential desk, Slater leaned back in the cushioned chair, thinking. "Okay," she finally said. "So where do we go from here? The date is critically tied to our agreement date with providing land to the other countries who have already purchased mirrors. In the event we can't come through, I'm sure that they will be asking for a speedy refund of their payments which will affect our country drastically and financially."

"We agree," said Killix, nodding. "That is why we believe that this *rest* will actually work to our advantage."

"How?" asked Slater.

"Yes how?" Secretary Hicks joined in.

"The Us have already stated they are unwilling to grant land expansion at this time," the General laid it out for them. "And they are not one to change their words. I've only known them to do it one time. In light of that and your desire...all of our desire for a diplomatic solution, we plan to still meet with them. They have given us an agreed upon meeting time, tomorrow, after their rest ends at twelve noon."

The General gestured over to Killix as he continued laying out the plan. "Mr. Killix and I, along with the Joint Chiefs and NATO command have worked up a plan where during their rest we can occupy the planet with the weapons necessary in the event that they do not change their minds and force becomes necessary."

The President let Maddox's words sink in, weighing all of her options. "So basically while they are sleeping, we move in with troops and weapons to demand their compliance in the event they are unwilling to change."

"I couldn't have said it better myself," Maddox said with a smile.

President Slater stared at the General and he stared back in utter silence. The two looked each other right in the eyes while Secretaries Killix and Hicks watched from the sidelines, in suspense. Finally the President swiveled in her chair so that they could only see the back of her head. "General Maddox," she said, "I want you to know something."

"Ma'am?" the General said, nervously shifting in his own chair while he waited for her to finish.

"General, I am not obligated to do what you want. I'm obligated to do what's best. I am aware that you already have the strike team and equipment on alert and ready at the requisition insertion points." She spun back around, looking right into the General's eyes. "So in light of that I can only assume two things. You either think you know me better than I think you know me, or you plan to execute your plan with or without my approval."

The General said nothing.

In the absence of a response, President Slater nodded and punched a button on the phone on her desk. "Send them in," she said.

Suddenly the door to the Oval Office swung open and six marines walked in along with four Secret Service agents, their eyes trained on the General Maddox and Killix who looked around, suddenly realizing what the President was threatening.

"So which one is it?" Slater asked.

"Madam President," Maddox said, his voice suddenly overflowing with humility, "I simply assumed you would do what is in the best interest of the country."

Slater stared him down for a long moment, then reached out and picked the executive order back up. Leafing through it, she paused briefly then reached for a pen. "Unfortunately for me and fortunately for you…" she said as she signed the order, "you're right. You have a go."

Killix's mouth broke into a wide smile. He nodded as he took the signed document from her. "It's the right thing to do."

"Really?" she asked, her voice flat and unsure. "Because I'm not convinced of what's right just yet. I am convinced of what's best and since the Us will be out of their rest in a little over twelve hours you had best get a move on."

General Maddox stood up from his seat. "Thank you, Ma'am. I'll get on it right away."

Maddox and Killix turned and headed for the door, but the Marines and the Secret Service agents refused to move. They stood their, blocking the two mens' paths of egress with stone faces.

"And General?" Slater called from behind them. Maddox turned back to her. "The next time you decide to cross me, I promise you those men won't leave empty handed. Are we clear?"

The General swallowed, nervous. "Crystal…" he nodded. "Ma'am. Was their anything else?"

President Slater thought about it. She only had one question. "Are you sure you haven't underestimated the Us, General?"

He nodded emphatically. "Positive. We've even spoken with Jergensen himself and he agrees that the Us will not resort to any type of physical violence in the event it comes to that. It is, how did he say it, against their code."

"And exactly how did you get Mr. Jergensen to even speak with you?" Slater wanted to know.

Maddox shrugged. "We offered him his job back."

Slater looked over to Killix. Killix flashed her a slight smile. "Not the exact same one," he explained. "But one similar."

As the President gave a slight nod, the Marines and Secret Service agents cleared their way and Maddox and Killix exited the office followed by the Marines and agents. President Slater stood staring at the closed door, still wondering if she had made the right decision.

Troyconics Astrophysics — Columbus, Ohio, U.S.A.
Earth Date: December 15th, 2015 — 8:07am EST

Calvin was just sitting down at his office desk, a coffee in one hand and a donut in the other, when Jerry burst through the door and slammed it shut behind him. "Calvin," he yelled frantically. "You've got to do something. We've got to warn them!"

"Good morning," Calvin said sarcastically. "Come right in, Jerry. Make yourself at home. Oh, you already have. You've only been back on the job for one day and you're already telling me what to do. Again! Welcome back."

"Screw the job!" Jerry yelled. "Why are you okay with this? Did you agree to this?"

"Jerry, I agreed to nothing. I told you, they simply asked me to get you here to meet with them. I found out their questions at the same time you did."

"They're going to attack the Us," Jerry said, as if it were the only thing in the world that mattered.

"Jerry, that's possible," Calvin agreed. "But it appears only if the Us do not agree to give permission for additional land expansion."

"I've got to warn them," Jerry said as he began to pace back and forth in Calvin's office. "Can you get me through the gate?"

Calvin shook his head. "You know better. All of Maddox's men are on this side and the other side of the insertion point. You're on a no-travel list. You'd be picked up and arrested in seconds."

Jerry stopped pacing long enough to think. Then he shook his head. "We can't just sit back and let this happen. We've got to let the ThreeofUs know."

"Jerry," Calvin said, looking his friend in the eye. "What makes you think they don't know?"

The Sanctuary of Us — Dabar
December 15th, 2015 12:00 P.M. (U.S.A EST)

While resting, the Us sanctuary was protected by a red glowing force field only visible to them. As they began to awaken from their slumber, however, the glow began to fade. Deep inside the castle, as each tubular door opened in turn and each of the Us stepped out of their resting chambers, the force field dissipated until they were all awake and the castle was left unprotected. This time tube filled hall extended longer than before.

Inside, the red light from the tubes filled the rooms with its eerie glow and the Onerous walked over to the ThreeofUs and the Fourofus. "And so it is time," he said.

ThreeofUs and FourofUs nodded. "And We are ready."

Outside the castle, Secretary Killix and Sergeant Barclay waited patiently at the gate, the General and his men hanging back a half mile or so unseen in the valley, so that the Us would not suspect any ill intent. Soon, after the glowing force field had disappeared, the gate opened of its own accord and the teenage form of the ThreeofUs stepped out into the sunlight.

"Okay," Killix breathed heavily, "here we go. Are we ready, Sergeant?"

Barclay nodded. "Everything is in place, Sir."

"Mr. Killix," the ThreeofUs said in greeting, "once more we meet." Then he turned to Barclay. "Sergeant Barclay, *once more only we meet again.*"

"The pleasure is mine," Killix said, ignoring the puzzling comment. "It's an honor to meet you. I've heard many things about you—" Barclay nudged Killix a reminder and the Secretary recovered somewhat gracefully. "—all. Many things about you all."

ThreeofUs nodded. "And We you." They stared at each other for a moment, then the ThreeofUs turned and motioned for the Secretary Killix to follow. "Join Us inside."

"I take it the rest proved refreshing," Killix commented as he was led through the mind-bending innards of the sanctuary.

"It has served its purpose," the ThreeofUs said. "In more ways than one."

Killix nodded and looked around, surveying the castle's interior. "Very unique place you have here," he said. "Very reminiscent of ancient castles on our planet."

"Yes, We know," the ThreeofUs said before gesturing to a lighted archway. "The terrace is an excellent site-seeing spot. And we should definitely take advantage of the beauty of the day. We both know how *quickly the weather can change.* Join Us?"

"Sure," Killix said, and followed the ThreeofUs out onto the balcony. The sky above was a vibrant blue with no clouds as it always was. Out in the distance, the ever-growing bustle of the United States of America on Dabar stood as a symbol of what the future could hold for the planet Dabar. Down below, they could see Sergeant Barclay waiting at the front of the Sanctuary.

"So," the ThreeofUs asked Killix as the two of them stood at the railing, looking down at Barclay, "what is it you would like to speak with Us of? General Maddox made your request for more land clear, as did We our concern regarding approving that request in light of recent developments in the territory you currently occupy."

"We understand and respect your or Us's concern," Killix said, trying to remain diplomatic. "However, it is my understanding that initially there was no restriction placed on where we could go and what we could do with the exception of the areas designated for Us."

"That is not entirely correct," the ThreeofUs said, "as we stated there would be stipulations based on what could be done in certain areas and materials used in those areas. Expansion of land was pending how the land you were given was properly utilized."

Killix nodded, choosing his words carefully. "On behalf of the President of the United States of America, we want to express our sincere gratitude for the camaraderie and free access that we have been allowed on Dabar. We sincerely appreciate it. It is our hope that we can continue to work together to a mutually beneficial purpose."

"Beneficial for whom?" the ThreeofUs asked, his face remaining neutral and expressionless. "The death of your citizens cannot certainly be deemed beneficial for them. Therefore We can only assume that the benefit you speak of is for your government and the financial gain of the sale of additional access points to other countries on your planet."

"ThreeofUs, please be reasonable," Killix implored the being. "We have enjoyed a working relationship to this point. It is our desire to grow that relationship. However, due to extenuating circumstances, a much more controlling interest is now necessary due to the economic ties and weight that Dabar now has in regard to the U.S. and other world economies. It would be unfortunate if we were forced to use non-diplomatic means to execute the necessities of this…transaction."

The ThreeofUs turned and stared at Secretary Killix. "Are We mistaken," he finally said, "in assuming that you, after friendly acceptance and open invitation to Our planet, are now considering an attempt at hostile takeover?"

Killix stared back, unafraid and unwilling to back down. "I would say at this point 'attempt' would be an understatement. The plan is already in motion."

Before ThreeofUs could respond, Killix waved down at Barclay, giving him the signal they had agreed on. On the ground, Barclay spoke a few mumbled words into his comm link. Within moments, Killix and the ThreeofUs saw tanks, armored vehicles, and thousands of ground troops spill over the crest of the valley, headed for the sanctuary. They continued forward, the tanks leveling their canons at the castle.

Killix shot the ThreeofUs a smug grin. They had cornered the Us exactly according to their plan and now, no matter what the being did, Dabar would belong to them. But the ThreeofUs seem unphased at all by the sight of the troops and the tanks. Instead, for the very first time, the ThreeofUs smiled.

Secretary Killix's smile, now faltered under the ThreeofUs sudden display of emotion. "Do you think that We are not aware of everything that has transpired since the first ejaculation of the sperm of your treachery?" the ThreeofUs asked, the smile never leaving his lips. "*All is known to Us,* and *only* what we deem necessary for you to know is what is known to you about Us."

Killix looked around nervously, suddenly wishing that the troops would hurry up and reach the sanctuary.

"You are under the misguided notion," the ThreeofUs went on, "that it is your world that sustains ours, when in truth your world depends on the literal underpinnings of all that is spoken and said by Us with *Substance*. If We but execute our will, your world will cease to exist. So in all seriousness, We implore you… *STAND DOWN!*"

They could now hear the rumbling treads of the tanks and the thunderous footsteps of the marching troops. Killix emboldened, reassured by the arrival of General Maddox and his men. "On the contrary," he said, his smile returning. "It is you who have underestimated 'us.' The U.S. We have people inside your very ranks that have revealed your threat level and trust me, it is not sufficient."

The ThreeofUs smiled again and took a step back from the Secretary. Down below, on the ground, Sergeant Barclay turned to find himself standing face to face with OneofUs. Startled, he jumped back, stumbling to the ground and looking up at the being.

"Foolish human," OneofUs said to him, shaking his head, "always confident in only what you can see. *Believe Us.* There are more that be with Us than be with you."

As soon as the words were out of his mouth, the rest of the Us conglomerate began to reveal themselves. Suddenly OneofUs became TwoofUs. TwoofUs multiplied into ThreeofUs and ThreeofUs into TwelveofUs. The Us continued to grow, multiplying on the ground below the terrace. Soon there were FiftyofUs and then a HundredofUs, then FiveHundredofUs and a ThousandofUs. Among their numbers also stood Ms. Jankowski, Max and Mack, Mark Grammar, Keltay, Efrat, Changyi and a host of other citizens from Earth whom had joined the Us. At the apparent end of the multiplication there were more Us on the ground than there were soldiers, completely outnumbering the General's troops.

Killix looked down from the balcony in fear and disbelief as he saw the

Us's numbers bloom. He yelled down to Barclay, "Prepare to engage!"

Still in shock from his encounter with the OneofUs, Barclay fumbled for his comm link and depressed the talk button. He relayed the message back to General Maddox and within seconds the United States troops had their weapons shouldered, ready for battle. The Us however, remained still.

"Mr. Secretary," the ThreeofUs said, "indulge Us please. Can you tell Us, do you know the date when Jerry Jergensen first came to Dabar?"

"Certainly," Killix said with contempt. "April 21st, 2014."

"And do you know what day it is today based on your calendar?"

"It is December 15, 2015."

The ThreeofUs smiled again. "You are incorrect," he said. "Today *is* April 20th, 2014."

And then, all at once, Secretary Killix disappeared. Sergeant Barclay disappeared. All of the tanks and troops and armored personnel carriers simply blinked out of existence. The buildings and streets and bustling residents of the United States on Dabar were suddenly gone, leaving no trace. The Dimensional mirror disappeared and the portal linking Earth to Dabar closed in a quiet zip. The only thing that remained was the original wilderness of Dabar as the other Us members on the ground drew their expanded likenesses back into themselves.. The valley looked just as it had when Jerry first arrived, lush and untouched. All was as it were.

The ThreeofUs turned away from the balcony's railing to find OneofUs standing behind him. "How many times will We allow this?" he asked the ThreeofUs. "As many as is necessary for them to get it right," ThreeofUs replied.

Troyconics Astrophysics — Columbus, Ohio, U.S.A.
Earth Date: April 21st, 2014 — 3:48am EST

"Finally," a soft voice muttered as the world was quietly ushered into a new era. Unbeknownst to the slumbering masses of Ohio, or the Japanese businessmen finishing up their work day in Tokyo, or the multitude of tourists in Paris who were just now eating breakfast and planning their day of site seeing, life on planet Earth had just undergone a fundamental change in course.

Jerry Jergensen stood over the central control console, the glow of the five separate monitors bathing his face in a pale light. He shook his head as he scanned the data read-out and confirmed what he was seeing. After years of failure he had finally succeeded in creating a device that pulled a parallel world close enough for a stable portal. The portal edges spun and glimmered ever so slightly as it simply stood there like an 8-foot horizontal disc in thin air beckoning to be entered. The two worlds now sitting ever so close. It looked as if you could simply step thru your doorway and then step back in. Everything was running smoothly, just the way it always had in his head. Only this time it was real; this time it worked.

He looked up from the monitors and once more marveled at his work. At the back of the otherwise empty laboratory, thirty feet away from the podium-like control console, Jerry's life's work whirred and spun and bent the known laws of nature. Red laser light bounced and reflected about the lab, a dazzling disco-ball blur of moving lines and dots transitioned from mirror to mirror and this self same light was the only thing holding the portal open. Jerry gulped a few ounces from a bottle of water as he walked closer to the portal. You could see straight thru to the other planet, which looked eerily similar to Earth, its hillside lush and green. Without thinking he threw water from the bottle towards the portal. As some drops fell to the lab floor, others entered the portal and fell into the grass. He paused as if waiting for something unexpected to happen. When nothing did, he quickly donned an environmental suit designed specifically for this day, the day he met with success and then clicked his helmet into place on the suit.

305

Jerry briefly thought of his wife Candy and his daughter Alexandria as he took a step towards the portal then stepped back.

"Jerry" he said to himself, "We didn't come all this way for nothing! The life insurance is paid up and Candy is still young, she can remarry." And with that selfish rationalization Jerry Jergensen closed his eyes and stepped into the portal. When his foot felt firm ground, he squinted open one eye then the other, like a child expecting to see the boogey man. Instead, he was surrounded by a perfectly blue sky and the green, lush countryside. He turned around to see the portal at his back and then looked back to the clear blue sky of the new planet and without thinking he uttered a single word. "Dabar!"

# # #

Dabar

dâbar, daw-bar'; a primitive root; perhaps properly, to arrange; but used figuratively (of words), to speak; rarely (in a destructive sense) to subdue:— answer, appoint, bid, command, commune, declare, destroy, give, name, promise, pronounce, rehearse, say, speak, be spokesman, subdue, talk, teach, tell, think, use (entreaties), utter

About the Author
Vincent H. Ivory

Unique, innovative and thought provoking are just some of the words used to describe Vincent Ivory and his literary work. A poet, author and screenwriter, Vincent provides us a unique twist on a view of the power of words in his debut novel *Word World*.

Born and raised in Detroit, Michigan, Vincent has always been a lover of books, writing and reading. At the tender age of six, his enthusiasm for reading was rewarded when he was removed from one of his regular first grade classes and placed in a unique program called Junior Great Books. In this venue, reading and vocabulary skills were honed as well as exposure to other great literature. "As a child I loved Curious George and the normal kid stuff but Rudyard Kipling's Rikki Tikki Tavi and classics like Beowulf and Grendel always captured my imagination." Vincent also grew up as an avid comic book reader. "We didn't have a lot of money so brand new comic books were out of my reach at the time. But in Detroit they had a three pack of comics in plastic and the covers were cut off so they were not worth as much and sold for cheap. I had stacks of those. I started working at age eleven sweeping and mopping floors at a hair salon with my Dad and a significant portion of that money went to comics. This trend continued well into my teens, twenties and beyond."

Today, Vincent is happily married, has three children and resides in a suburb of Columbus, OH. He works on his books, screenplays and marketing every single day (sometimes to his wife's dismay). Vincent has always loved and written poetry throughout his childhood and adult life. "Writing is very therapeutic for me and I believe it can be that way for everyone. Never say I can't write. That is simply not true. You can write. You may not write up to other people's expectations or standards but you can write and you should. Journal your life, your feelings and experiences. Writing is a great way to express joy and relieve anger or unhealthy emotions and thoughts because you can then look at how you are feeling and thinking. You literally release those thoughts and emotions onto paper instead of keeping them in."

Contact Vincent at going2dabar@gmail.com or visit vincentivory.com or like us on https://www.facebook.com/go2wordworld

The Made it Happen Team

I wanted to dedicate this section to sincerely thank everyone who were a part of the Make it Happen Team!!!

Dominique Boyd
Julius & Sandra Booker
Briana Booker
Marianne Coyne
Sylvia Ivory
James Ivory
Ralph Johnson
Harvey Austin
Andre O'Neal
4 Anonymous Persons who requested that they remain so
(U know who U R)

Simple Rhyme Poetry or Other books

by

Vincent H. Ivory

Love Letters to Hollywood Vol. 1 & Other Not So Randumb Thoughts

In Love Letters to Hollywood, Vincent allows us a rhythmic peek at some of his caring and many times comedic insights and views on a select group of Hollywood's biggest and upcoming names. The letters, written with great candor, are filled with observations, opinions, little known facts and origins some of which reveal immediately who the celebrity is, while others may take a bit of research. In either case the letters make for a fun and reflective time of literary indulgence that is rare. Opposing all naysayers, Vincent continues to disperse the joy of rhyme poetry with an unrelenting resilience encouraging readers and writers everywhere to do the same.

Do Pantyhose Lie?

Do Pantyhose Lie? is an intimate collection of poetry about life, love, emotions, creativity, sensuality and spirituality. With an unorthodox yet contemporary style, Vincent Ivory engages the timeless questions of purpose and passion while admiring the tangible and intangible beauty of mature women in pieces like Women & Wine. Every reader will find themselves in at least one of the creations as they discover a collection of poetry that definitively makes you go ... hmmm?

Fanmail: Poetry Inspired by the Characters
Of Marvel Comics

FANMAIL is a unique and innovative poetic commentary and tribute to many of the superheroes we have grown up with and admire. Fans young and old will relate to the gritty yet truthful comments in a simple rhyme format.

To find out more or order visit *vincentivory.com*

What's Next?

I truly hope you enjoyed Word World as much as I enjoyed writing it. Words are powerful. There is a popular saying, "Sticks and stones may break your bones but words can never hurt me." Nothing could be further from the truth. Words can heal and words can hurt. Words can have lasting effects for hundreds or even thousands of years by typecasting or profiling. So I encourage you to watch your words, towards yourself, your family, your friends and loved ones and especially children. One "Your so stupid!" can cut deeper than the sharpest blade and affect self esteem for years to come.

With that said, there are so many ways we can go with this story as this was simply an origin story, an introduction. There is so much more we still don't know about The Us as well as the full potential of what can be done on the parallel planet Dabar. So what would YOU like to see in the next installment?

What are your questions or thoughts? Did you ever think or ask yourself any of the following questions:

Where are the animals?

How come there is no Sun and where is the light coming from?

How come there is no mention of bodies of water of Dabar except one?

How young can people become on Dabar?

Was the ThreeofUs responsible for Jerry's discovery?

So tell me what would you like to see? Reach out to us at vincentivory.com or going2dabar@gmail.com.

Thanks for being a partner in What's Next! All the best to you!!!

Sincerely,

Vincent H. Ivory
"*Believing is Seeing*!!!"